SpringerBriefs in Energy

More information about this series at http://www.springer.com/series/8903

Antonio Moñino · Encarnación Medina-López
Rafael J. Bergillos · María Clavero
Alistair Borthwick · Miguel Ortega-Sánchez

Thermodynamics and Morphodynamics in Wave Energy

 Springer

Antonio Moñino
Andalusian Institute for Earth
  System Research
University of Granada
Granada
Spain

María Clavero
Andalusian Institute for Earth
  System Research
University of Granada
Granada
Spain

Encarnación Medina-López
School of Engineering
University of Edinburgh
Edinburgh
UK

Alistair Borthwick
Institute for Energy Systems
University of Edinburgh
Edinburgh
UK

Rafael J. Bergillos
Andalusian Institute for Earth
  System Research
University of Granada
Granada
Spain

Miguel Ortega-Sánchez
Andalusian Institute for Earth
  System Research
University of Granada
Granada
Spain

ISSN 2191-5520   ISSN 2191-5539   (electronic)
SpringerBriefs in Energy
ISBN 978-3-319-90700-0   ISBN 978-3-319-90701-7   (eBook)
https://doi.org/10.1007/978-3-319-90701-7

Library of Congress Control Number: 2018939460

Printed on acid-free paper

This Springer imprint is published by the registered company Springer International Publishing AG part of Springer Nature
The registered company address is: Gewerbestrasse 11, 6330 Cham, Switzerland

Some may never live, but the crazy never die.

Hunter S. Thompson

Preface

This Springer Briefs volume on the thermodynamics and morphodynamics in wave energy aims to present some latest achievements in this field for people who are interested to learn more how this type of factors influences both wave energy extraction and their impacts on the coast. It is intended for engineers, researchers and postgraduate students in the areas of Renewable Energies, Coastal Engineering, Physics and Sustainability who are concerned with the evolution and sustainable management of the human environments.

The book focuses on two key factors concerning the global performance of oscillating water column wave energy converters, namely the governing thermodynamic factors influencing the turbine efficiency and the morphodynamic processes involved in the mutual interference between the propagating waves and the converter. On the one hand, the conditions of the air mixture inside the chamber and the surrounding atmospheric system set the scenario for the compression/expansion thermodynamic process inside the wave energy converter during the wave cycle. On the other hand, there is a mutual influence between the device interaction with the impinging waves and the seabed evolution, which in turn affects lately the device performance in such a closed loop working cycle. Chapter 1 introduces a state of the art of the wave energy extraction, focused on the oscillating water column technology as one of the most widely considered for successful design, deployment and farming. Chapter 2 presents the first results that help to understand one of the possible reasons for the low efficiency in the oscillating water column devices, that is, the presence of an air–water vapour mixture as the working fluid inside the chamber and its behaviour as a real gas rather than as an ideal one. Chapter 3 deepens in the concept of real gas applied to the compression/expansion thermodynamic process inside the chamber. Once the basic facts have been presented and the basic theoretical concepts have been developed, Chap. 4 is devoted to provide with a proposed methodology for the simulation of the oscillating water column performance in which key factors involving the linear turbine response and the wave–device interaction are implemented. Chapter 5 sets the guidelines for the study of the wave–device interaction and the mutual effects on the morphological seabed evolution and the overall device efficiency. Finally, and once the all the basic concepts have been developed throughout the previous chapters, the research is completed in Chap. 6 with a study on how wave energy converter farms can be used not only for a regular ocean energy farming policy, but also and not less important, to preserve the coastal environment and the morphodynamic evolution between sustainable limits.

The improvement in the oscillating water column technology has to account for the eventually low value of efficiency and its causes, balancing it by reducing costs of design, manufacturing, deployment, operation and replacement. Although that technology has by now a long way to go in terms of those issues yet to be improved, the context is not different from the first steps taken years ago by other clean energy sources such as wind or solar, which now represent a standard in terms of engineering and technology applied to clean energy. Indeed, it is not a question of durability and/or maximum efficiency. It is a matter of fixing appropriately the expected values of efficiency and production, and then adapt design, materials, deployment and maintenance according to those predictions, so that the net balance be positive. With all the theoretical background and the information provided by numerical and experimental simulations, the researchers in this project consider that the moment is now to start out new guidelines for cost-effective design and deployment of easy-to-manufacture, competitive devices. Although this book is mainly focused on those results and the corresponding publications, a large list of some of the most relevant and up-to-date references is provided in each chapter. We apologize for any errors that may be found in the book, despite our efforts to eliminate them.

Finally, we would like to highlight that the study of these marine renewable energies began circa 2006, in the context of some specific research technological projects which, as alternate technical solutions, faced the wave energy extraction as one possible option to overcome power supply issues. The authors are grateful to the members of the Environmental Fluid Dynamics Group of the University of Granada, particularly to Pedro J. Magaña for his assistance in processing the text and figures. We are indebted to Asunción Baquerizo and Miguel A. Losada for their valuable contributions to the book. We would also like to thank the support provided by the Andalusian Regional Government (Project P09-TEP-4630).

Granada, Spain   Antonio Moñino
March 2018   Encarnación Medina-López
                                                                                          Rafael J. Bergillos
                                                                                              María Clavero
                                                                                          Alistair Borthwick
                                                                              Miguel Ortega-Sánchez

Contents

Symbols and Greek

Chapter 2

Symbols

A_t	Turbine cross-sectional area
A_a	Turbine blades area
$\mathbb{B}$	Second virial coefficient
$\tilde{B}$	Oscillation damping coefficient
c	Turbine blades chord length
C_g	Wave group celerity
C_p	Specific heat for the ideal gas, constant pressure
C_{pa}	Specific heat for dry air, constant pressure
C_{pg}	Specific heat for the real gas, constant pressure
C_{pv}	Specific heat for water vapour, constant pressure
C_s	Speed of sound in air
$\mathbb{C}$	Third virial coefficient
$\tilde{C}$	Restitution coefficient
D_t	Turbine diameter
e	Vapour pressure
e_s	Saturation vapour pressure
$e_0 = 0.611 \text{ kPa}$	Saturation vapour pressure at 273 K
f_0, f_1, f_2	Temperature correlation functions
g	Gravity acceleration
H	Enthalpy for the ideal gas; wave height
H_g	Enthalpy for the real gas
k	Wave number
K	Turbine characteristic number
$L = 2.5 \times 10^6 \text{ J/kg}$	Latent vaporization heat
m_a	Mass of dry air

m_v	Mass of vapour
MW_a	Molar weight of dry air
MW_v	Molar weight of water vapour
N	Turbine rotation velocity
p_a	Pressure of dry air
p_c	Critical pressure
p_{ca}	Critical pressure for dry air
p_{cv}	Critical pressure for water vapour
p_g	Pressure of air–water vapour mixture
p_{in}	Pressure at the turbine inlet
p_{out}	Pressure at the turbine outlet
p_{owc}	Pressure (manometric) inside the OWC chamber
p_r	Reduced pressure, $p_r = p/p_c$
p_0	External thermodynamic reference pressure
p_2	Total thermodynamic pressure inside the OWC
$\widehat{P}_{owc}$	Complex pressure amplitude in the OWC
Q_{owc}	Volumetric flow rate at the turbine inlet
Q_{owc}^m	Mass flow at the turbine inlet
Q_T^m	Mass flow through the turbine
$\widehat{Q}_{owc}$	Complex OWC air flow amplitude
r	Absolute humidity
$R_a = 286.7 \text{ J/K kg}$	Dry air constant
R_{ext}	Turbine external radius
R_{int}	Turbine internal radius
$R_{med} = \frac{R_{int} + R_{ext}}{2}$	Turbine mean radius
R_g	Real gas constant
R_{owc}	OWC radius
$R_v = 461 \text{ J/K kg}$	Water vapour gas constant
$R_0 = 8.31 \text{ J/K mol}$	Universal gas constant
RH	Relative humidity
S_{in}	Turbine inlet section
S_{out}	Turbine outlet section
T	Temperature
$T_{ca} = 132 \text{ K}$	Critical temperature for dry air
$T_{cv} = 647 \text{ K}$	Critical temperature for water vapour
T_{in}	Temperature at the turbine inlet
T_r	Reduced temperature, $T_r = T/T_c$
$T_0 = 273 \text{ K}$	Reference temperature
U	Blade circumferential velocity
V_g	Gas volume
V_{owc}	Air volume inside the chamber
V_x	Turbine axial velocity
w	Air velocity relative to the turbine
$\overline{W}$	Average power
$\mathbb{Z}$	Compressibility factor per gas mole

Greek

β	Coefficient depending on turbine characteristics
$\gamma = c_p/c_v = 1.4$	Heat capacity ratio for air
$\tilde{\Gamma}$	Coefficient for diffraction problem
$\varepsilon = R_a/R_v$	Ratio of gas constants for dry air and water vapour
χ	Coefficient for turbine characteristics
χ_{mol}	Molar fraction of vapour in dry air
ρ_a	Dry air density
ρ_0	Air density outside the chamber
ρ_2	Air density inside the chamber
σ	Turbine solidity
ω	Acentric factor

Chapter 3

Symbols

B	Second virial coefficient
$B' = B/M$	Second virial coefficient per gas mole
C_p	Specific heat for the real gas, constant pressure
C_p^*	Specific heat for the ideal gas, constant pressure
$\tilde{C}_p$	Non-dimensional real gas specific heat at constant pressure
C_s^*	Speed of sound in dry air
C_s	Speed of sound in a real gas
C_v	Specific heat for the real gas, constant volume
C_v^*	Specific heat for the ideal gas, constant volume
$C_y = T\left(\frac{\partial S}{\partial T}\right)_y$	Specific heat for constant variable y
e	Vapour pressure
e_s	Saturation vapour pressure
$e_0 = 0.611$ kPa	Saturation vapour pressure at 273 K
f_0, f_1, f_2	Temperature correlation functions
g	Gravity acceleration
h^*	Enthalpy per mole unit for the ideal gas
h	Enthalpy per mole unit for the real gas
$\tilde{h}$	Non-dimensional enthalpy per mole unit for the real gas
H^*	Enthalpy for the ideal gas
H	Enthalpy for the real gas
k_T	Isothermal compressibility factor
$L = 2.5 \times 10^6$ J/kg	Latent vaporization heat

$m = \dfrac{C_y - C_p}{C_y - C_v}$	Index m
m	Mass
M	Molecular weight
n	General polytropic index
P_{PTO}	Power available to the turbine
P_w	Power in the OWC chamber
p	Pressure
p_c	Critical pressure for water vapour, 220.89×10^5 Pa
p_g	Pressure of air–water vapour mixture
p_{in}	Pressure at the turbine inlet
p_{out}	Pressure at the turbine outlet
p_r	Reduced pressure, $p_r = p/p_c$
p_0	Reference thermodynamic pressure
Q	Heat
Q_w	Flow rate driven by water surface inside the OWC chamber
Q_p	Air flow rate through the OWC turbine
r	Absolute humidity
$R_a = 286.7$ J/K kg	Dry air constant
$R_g = R_0/M$	Real gas constant
$R_v = 461$ J/K kg	Water vapour gas constant
$R_0 = 8.31$ J/K mol	Universal gas constant
Rg	Real gas non-dimensional number
RH	Relative humidity
s^*	Molar entropy for the ideal gas
s	Molar entropy for the real gas
S_{in}	Turbine inlet section
S_{out}	Turbine outlet section
T	Temperature
T	Period
T_c	Critical temperature for water vapour, 647 K
T_g	Temperature at the turbine inlet
T_{out}	Temperature at the turbine outlet
T_r	Reduced temperature, $T_r = T/T_c$
$T_0 = 273$ K	Reference temperature
t	Time
u^*	Internal energy of a system per mole unit for the ideal gas
u	Internal energy of a system per mole unit for the real gas
$\tilde{u}$	Non-dimensional internal energy per mole unit for the real gas
U	Internal energy of a system
U_{in}	Air velocity at the turbine inlet
U_{out}	Air velocity at the turbine outlet
$v = V/N$	Volume per mole unit
V	Volume

| x | Main horizontal direction |
| Z | Compressibility factor per gas mole |

Greek

Δs^*	Ideal gas entropy per mole unit variation
Δs	Real gas entropy per mole unit variation
$\Delta \tilde{s}$	Non-dimensional real gas entropy per mole unit variation
$\Delta \psi$	Phase variation between ideal and real airflows
δH	Deviation from the ideal enthalpy
η	Efficiency
$\gamma = C_p / C_v = 1.4$	Adiabatic index for an ideal gas. Heat capacity ratio for air
$\varepsilon = R_a / R_v$	Ratio of gas constants for dry air and water vapour
χ_{mol}	Molar fraction of vapour in dry air
μ^*	Chemical potential of an ideal gas
μ	Chemical potential of a real gas
Ψ	Displacement from equilibrium position
$\rho = N/V$	Molar density
ρ_a	Dry air density
ρ_g	Real gas density
ρ_0	Reference air density
ω	Acentric factor

Superscript Index

| * | Variable for ideal gas |

Chapter 4

Symbols

a	Axial induction factor
A	Cross section
A	Actuator disk cross section
A	Wave amplitude
c	Turbine chord length
C_P	Power coefficient

C_T	Thrust coefficient
C_2	Inertial resistance factor
d	Turbine porosity
D_t	Turbine diameter
h	Water depth
h	Hub-to-tip ratio
h_{owc}	Water depth at OWC location
h_{ref}	Reference water depth
H	Wave height
k	Wave number
K_L	Loss coefficient
K_L'	Porous zone equivalent loss coefficient
$\dot{m}$	Mass flow
N_b	Number of blades
P	Power
R_h	Hub radius
R_t	Tip radius
S	Paddle stroke
S_{ag}	Net air gap section
S_i	Source term for the momentum equation
S_{owc}	OWC horizontal section
$S_{turbine}$	Turbine section
S_0	Initial paddle stroke
t	Simulation time
T	Wave period
T	Thrust
U	Velocity
U_{ag}	Air gap velocity
U_d	Induction velocity
U_{owc}	Vertical water surface velocity inside OWC
U_{paddle}	Paddle displacement velocity
U_{ring}	Velocity in a ring-like section (full section with central hub, without blades)
v	Velocity
V_{owc}	Air volume inside the chamber
V_w	Velocity at the inlet of the stream tube
x_{paddle}	Paddle displacement along horizontal axis

Greek

α Permeability
η Surface elevation
δ Porous zone thickness
Δp Pressure drop
μ Dynamic viscosity
ρ Density
σ Turbine solidity

Chapter 1
Introduction

Abstract Ocean waves offer a field worth of energetic use. Power from waves impinging the coast worldwide can be estimated in 106 MW, reaching 107 MW if that power is farmed off-shore, (Cruz in Ocean Wave Energy. Springer, 2008, [6]), (Falnes in Mar Estruct 20:185–201, 2007, [13]). Up to now, a long road has been travelled regarding the essentials in the start up of basic concepts and technologies, leading to an open field for a competitive use of the ocean energy resource. Once the world wide ground has been set ready for the urgent need of a better environmental use of natural resources, and once the main technical limitations, gaps and barriers and main economical, social and political issues have been identified, the time is now to take a step further and start to advance in the enhancement of technologies that make the concept of ocean energy usage become a sustainable reality. This book provides with a contribution along a path open by others, in which some advances and improvements in the field of wind wave energy resource conversion are presented, as a contribution aimed to a more efficient usage of clean and sustainable resources in general and to OWC more specifically.

1.1 The Role of Ocean Energy

Ocean energy offers future opportunities in the sector of renewable energy technology, and presently has the highest annual growth rate in its field, [24]. The ocean energy resource offers an attractive field for stakeholders in the renewable energy sector. There remain, however, many technical and financial challenges when trying to bridge the gap between research and technological deployment of full scale prototypes ready to supply the electrical grid. As a reference frame, in the case of Europe the SI Ocean Project (http://www.si-ocean.eu) was conceived to provide with a framework in which aspects concerning ocean energy extraction, both wave and tidal sectors, could be discussed, yet the foundations for a mid time scale deployment improvement and market growth were settled [1, 2]. The conclusions of that project can be made extensive to other countries.

The costs of wave energy extraction essentially depend on three factors [1]: the available resource, the technical features of the energy converter and the operational

© The Author(s) 2018
A. Moñino et al., *Thermodynamics and Morphodynamics in Wave Energy*,
SpringerBriefs in Energy, https://doi.org/10.1007/978-3-319-90701-7_1

time span of the device relative to its lifetime. For the renewable energy projects to be successfully conducted, not only scientific, technical and economic problems have to be solved. In addition, social acceptance might also face a risk of turning around into opposition by several concerned parties according to what it can be explained in simple terms by means of the NIMBY (*Not In My BackYard*) model [19]. Of all factors concerning the change in acceptance, environmental issues play a major role [18].

Summarizing, to bring forward such technology, research has to be undertaken aimed to specific objectives: from scientific advances concerning the governing principles, to technological issues involved in field deployment of single prototypes, and environmental policies envisaging the use of arrays of wave energy converters (hereinafter WECs) for both protection and clean energy production.

1.2 The Oscillating Water Column Energy Converter: A State of the Art Review

A variety of wave energy converters has been devised during the last two centuries, leading to the development of scientific and technical knowledge, focused on a set of several specific devices [6, 10]. Among all those devices, the oscillating water column (hereinafter OWC) is one to which more research and interest have been devoted, and it is one of the few devices to have been tested at full scale under prototype conditions [6]. The OWC basically consists of a partially submerged chamber opened to the sea at the bottom, and a power take off system (hereinafter PTO) to transform the wave–induced pneumatic energy inside the air chamber into electrical energy. The PTO is usually a Wells turbine, whose reversibility and linear response between air flow discharge and pressure are the main features.

Much research attention has focused the effect of key variables and parameters on the performance of OWCs in delivering power to the grid. OWCs are found to be sensitive to wave interaction, power take-off (PTO) performance, turbine and air flow dynamics, the effect of random sea states, control strategies, and optimization. Extensive research has been devoted in the past decades to solve the OWC problem. Starting with the classical theoretical formulation of the radiation–diffraction problem for a wave energy converter in its standard form, [11, 12, 34] and other authors have derived analytical solutions focusing on different factors concerning the OWC configuration, [9, 28–30]. Certain studies have gone further in considering the implementation of technical aspects regarding the Wells turbine as the power take–off system, [14, 15, 33, 35]. Other contributions identified mechanical improvements to the PTO device, [23], and devised methods to handle the random nature of sea waves based on signal control, [22]. Researchers have proposed ways to maximize turbine efficiency, based on direct control of either the turbine rotation dynamics, [8, 9, 15]), or the air flow through the turbine shaft, [8].

Numerical modelling has facilitated further research investigations into OWC optimization. Numerical flumes has been set ready to examine the effect of chamber geometry and turbine parameters on OWC performance, [38]. Linear and Non–linear system identification methods have been applied to study the multi–frequency excitation of water within an OWC chamber, [7, 16]. Numerical simulations have been conducted to learn on the OWC performance under different turbine damping values and tidal level variations, [25–27]. Recent studies have investigated the maximization of energy extraction according to the time scale considered (sea state, season, year), [21]. Further steps have also been taken to improve OWC design not only in terms of technological features, but also in the minimization of environmental impacts and subsequent restoration costs. Research has been carried out on the advantages of using OWC devices as coastline protection elements by [3] and [31], and for enhanced energy dissipation in breakwater structures by [17].

1.3   Open Lines in OWC Research: Looking to the Future

During the last decades, OWC power plants have been deployed worldwide: Portugal (Pico), Spain (Mutriku), Scotland (LIMPET), Norway (Tofftestallen), Australia (Port Kembla), Japan (Sakata, Kujukuri, Sanze, Niigata), India (Vizhinjam) and China (Shanwei, Dawanshan), among others. Those experiences and projects provide valuable data regarding performance and efficiency whenever they are available, [20, 39], revealing that the total performance of wave to electric energy conversion is about 10% in reported cases, even if predictions estimated total performances ranging from roughly 40–70%. This is especially relevant in the case of deployment of devices in the less energetic wave climates such as the Mediterranean Sea [37]. In such cases, it is essential an efficient performance of the converter, noting that [21] have found that optimal device performance can be attained in waves of peak periods about $\simeq 6$–7 s.

For the OWC technology to be competitive, not only design and deployment costs has to be reduced. It is also required an improvement in the global efficiency of the energy converter. The global efficiency represents the balance between the hydrodynamic efficiency (the ratio between compression/expansion work and wave energy) and the turbine efficiency (the ratio between turbine work and pneumatic work). In fact, control strategies oriented to maximize the global efficiency should not always be based on increasing the hydrodynamic efficiency, due to counter effects on the turbine efficiency [35].

One key factor in the research advance for the study of the turbine efficiency is the Thermodynamics of the gas phase inside the chamber. The formulation of the OWC problem usually assumes the gas phase as dry air, henceforth deemed as an ideal gas undergoing an adiabatic polytropic process of compression/expansion. The air process inside the chamber has been studied by several authors so far, [8, 36, 42], but no further questioning on either the ideal behaviour of the gas phase or the thermal isolation of the process is assumed. While it is reasonable to accept the adiabatic nature of the air cycle inside the chamber, some considerations on the constituents of the gas phase, i.e. a mixture of dry air and water vapour, might lead to a more realistic representation of the thermodynamics inside the OWC chamber, and in fact of the actual working conditions in a full scale prototype. Deviations from the ideal gas performance have been experimentally observed in a stationary two–phase air–water vapour flow through an OWC chamber model as described in this book. Moreover, the implementation of a real gas formulation in the OWC thermodynamics, based on the virial expansion—Kammerlingh–Onnes expanssion—, [32, 40, 41], reveals a diminishing in the global efficiency of the OWC, which might help explaining the differences pointed out above between predicted global efficiency and reported values.

Last, but not least, attention has to be paid to the different interactions with the coastal environment. In that sense the performance optimization of OWC energy converters at a given location might also be conditioned by induced changes in the hydrodynamics and their consequences. Namely, changes in the seabed due to variations in sediment transport could lead to geometrical modifications in the gap below the OWC chamber, leading to variations in the power efficiency. Even if those changes evolve in a mid–term scale, their effects on the day–to–day operation and their consequences on the lifetime energy production might not be negligible. But the importance of those facts not only goes in the direction of cost reduction and efficient investment; as it has been pointed out above, the acceptance of a project of this kind requires the fundamental environmental aspects to be fixed. In particular, facts concerning coastal dynamics have to be outlined and solved. As a first approach to the question, authors have recently came up with the idea of using wave energy converters as energy dissipation enhancement in breakwaters [17]. They have also been proposed as coastal defence elements, accounting for the fact that both the wave energy absorbed by the array and the dissipation induced by the presence of the obstacle reduce the energy impinging the coast [3–5, 31].

References

1. A., V.V.A.: Ocean energy: cost of energy and cost reduction opportunities. Technical Report, Strategic Initiative for Ocean Energy (2013). http://www.si-ocean.eu/en/Home/Home/
2. A., V.V.A.: Ocean energy: state of the art. Technical Report, Strategic Initiative for Ocean Energy (2013). http://www.si-ocean.eu/en/Home/Home/
3. Abanades, J., Greaves, D., Iglesias, G.: Wave farm impact on the beach profile: a case study. Coast. Eng. **86**, 36–44 (2014)
4. Abanades, J., Greaves, D., Iglesias, G.: Coastal defence through wave farms. Coast. Eng. **91**, 299–307 (2014)
5. Abanades, J., Greaves, D., Iglesias, G.: Coastal defence using wave farms: the role of farm-to-coast distance. Renew. Energy **75**, 572–582 (2015)
6. Cruz, J.: Ocean Wave Energy. Current Status and Future Perspectives. Springer (2008). ISBN 978-3-540-74894-6
7. Davidson, J., Giorgi, S., Ringwood, V.: Linear parametric hydrodynamic models for Ocean wave energy converters identified from numerical wave tank experiments. Ocean Eng. **103**, 31–39 (2015)

8. de O Falcão, A.F., Justino, P.A.P.: OWC wave energy devices with air flow control. Ocean Eng. **26**, 1275–1295 (1999)

9. de O Falcão, A.F.: Control of an oscillating wave energy plant for maximum energy production. Appl. Ocean Res. **24**, 73–82 (2002)

10. de O Falcão, A.F.: Wave energy utilization: a review of technologies. Renew. Sustain. Energy Rev. **14**, 899–918 (2010)

11. Evans, D.V.: Wave power absorption by systems of oscillating pressure distributions. J. Fluid Mech. **114**, 481–499 (1982)

12. Evans, D.V., Porter, R.: Hydrodynamic characteristics of an oscillating water column device. Appl. Ocean Res. **17**, 155–164 (1995)

13. Falnes, J.: A review of wave-energy extraction. Mar. Estruct. **20**, 185–201 (2007)

14. Gato, L.M.C., de O Falcão, A.F.: On the theory of the wells turbine. Trans. ASME **106**, 628–633 (1984)

15. Gato, L.M.C., de O Falcão, A.F.: Aerodynamics of the wells turbine: control by swinging rotor blades. Int. J. Mech. Sci. **31**(6), 425–434 (1989)

16. Gkikas, G.D., Athanassoulis, D.A.: Development of a novel Non-linear system identification scheme for the pressure fluctuation inside an oscillating water column wave energy converter part I: theoretical background and harmonic excitation case. Ocean Eng. **80**, 84–89 (2014)

17. He, F., Huang, Z.: Hydrodynamic performance of pile-supported OWC-type structures as breakwaters: an experimental study. Ocean Eng. **88**, 618–626 (1984)

18. Heras-Saizarbitoria, I., Zamanillo, I., Laskurain, I.: Hydrodynamic performance of pile-supported OWC-type structures as breakwaters: an experimental study. Renew. Sustain. Energy Rev. **27**, 515–524 (2013)

19. Hitzeroth, M., Megerle, A.: Renewable energy projects: acceptance risks and their management. Renew. Sustain. Energy Rev. **27**, 576–584 (2013)

20. Ibarra-Berastegui, G., Sáenz, J., Ulazia, A., Serras, P., Esnaola, G., García-Soto, C.: Electricity production, capacity factor, and plant efficiency index at the mutriku wave farm (2014–2016). Ocean Eng. **147**, 20–29 (2018)

21. Jalón, M.L., Baquerizo, A., Losada, M.A.: Optimization at different time scales for the design and management of an oscillating water column system. Energy **95**, 110–123 (2016)

22. Jefferys, E.R.: Simulation of wave power devices. Appl. Ocean Res. **6**(1), 31–39 (1984)

23. Korde, U.A.: A power take-off mechanism for maximizing the performance of an oscillating water: column wave energy device. Promotion policies for renewable energy and their effects in taiwan. Appl. Ocean Res. **13**, 75–81 (1991)

24. Krewitt, W., Nienhaus, K., Kleßmann, C., Capone, C., Stricker, E., Graus, W.: Role and potential of renewable energy and energy efficiency for global energy supply. Dec. Report No.: (UBA-FB) 001323/E. Technical Report. Federal Environment Agency (Umweltbundesamt), Dessau-Roßlau (2009)

25. López, I., Iglesias, G.: Efficiency of owc wave energy converters: a virtual laboratory. Appl. Ocean Res. **44**, 63–70 (2014)

26. López, I., Pereiras, B., Castro, A., Iglesias, G.: Optimisation of turbine-induced damping for an owc wave energy converter using a rans-vof numerical model. Appl. Energy **127**, 105–114 (2014)

27. López, I., Pereiras, B., Castro, F., Iglesias, G.: Performance of owc wave energy converters: influence of turbine damping and tidal variability. Int. J. Energy Res. **34**(4), 472–483 (2015)

28. Lovas, S., Mei, C.C., Liu, Y.: Oscillating water column at a coastal corner for wave power extraction. Appl. Ocean Res. **32**, 267–283 (2010)

29. Martins-Rivas, H., Mei, C.C.: Wave power extraction from an oscillating water column along a straight coast. Ocean Eng. **36**, 426–433 (2009)

30. Martins-Rivas, H., Mei, C.C.: Wave power extraction from an oscillating water column at the tip of a breakwater. J. Fluid Mech. **626**, 395–414 (2009)

31. Mendoza, E., Silva, R., Zanuttigh, B., Angelelli, E., Andersen, T.L., Martinelli, L., Nørgaard, J.Q.H., Ruol, P.: Beach response to wave energy converter farms acting as coastal defence. Coast. Eng. **87**, 97–111 (2014)

32. Prausnitz, J., Lichtenthaler, R., Gomes de Azevedo, E.: Molecular Thermodynamics of Fluid–Phase Equilibria. Prentice–Hall (1999). ISBN 0-13-977745-8
33. Raghunathan, S.: The wells turbine for wave energy conversion. Prog. Aerosp. Sci. **31**, 335–386 (1995)
34. Sarmento, A.J., de O Falcão, A.F.: Wave generation by an oscillating surface-pressure and its application in wave-energy extraction. J. Fluid Mech. **150**, 467–485 (1985)
35. Sarmento, A., Gato, L., de O Falcão, A.F.: Turbine-controlled wave energy absorption by oscillating water column devices. Ocean Eng. **17**(5), 481–497 (1990)
36. Sheng, W., Alcorn, R., Lewis, S.: On thermodynamics of primary energy conversion of OWC wave energy converters. Renew. Sustain. Energy **5**, 023,105–1–17 (2013)
37. Stefanakos, C.N., Athanassoulis, G.S., Cavaleri, L., Bertotti, L., Lefevre, J.M.: Wind and wave climatology of the mediterranean sea. Part II: Wave statistics. In: Fourteenth International Offshore and Polar Engineering Conference, Toulon, France (2004)
38. Teixeira, P., Davyt, D., Didier, E., Ramalhais, R.: Numerical simulation of an oscillating water column device using a code based on navier-stokes equations. Energy **61**, 513–530 (2013)
39. Trust, T.C.: Oscillating water column wave energy converter evaluation report. Technical Report, Marine Energy Challenge (2005)
40. Tsonopoulos, C., Heidman, J.L.: From the virial to the cubic equation of state. Fluid Phase Equilib. **57**, 261–276 (1990)
41. Wisniak, J.: Eike Kamerlingh-the virial equation of state. Indian J. Chem. Technol. **10**, 564–572 (2003)
42. Zhang, Y., Zou, Q.P., Greaves, D.: Air water two-phase flow modelling of hydrodynamic performance of an oscillating water column device. Renew. Energy **49**, 159–170 (2012)

Chapter 2
A Real Gas Model for Oscillating Water Column Performance

Abstract Oscillating Water Column (OWC) are devices for wave energy extraction equipped with turbines for energy conversion. The purpose of the present chapter is to study the thermodynamic of a real gas flow through the turbine and its differences with respect to the ideal gas hypothesis, with the final goal to be applied to OWC systems. The effect of moisture in the air chamber of the OWC entails variations on the atmospheric conditions near the turbine, modifying its performance and efficiency. In this chapter the influence of humid air in the performance of the turbine is studied. Experimental work is carried out and a real gas model is asserted, in order to take a first approach to quantify the extent of influence of the air-water vapour mixture in the turbine performance. The application of a real gas model and the experimental study confirmed the deviations of the turbine performance from the expected values depending on flow rate, moisture and temperature.

2.1 Objective

The objective of this chapter is to set the fundamentals of air turbine operation under changes in air density induced by variation of moisture content in steady state and their influence on the expected power output from OWC systems. A real gas model is proposed and experiments are conducted to measure the extent of influence of moisture on turbine performance. Although the method is usually applied in different fields from that of sea waves energy extraction, the fundamentals of real gases could bring up a new scope in the real working conditions of OWC devices.

2.2 Fundamentals of the Air Exchange Through the Turbine in OWC Devices

In an OWC system, the turbine is used as power take off system for pneumatic into electric energy conversion to be supplied to the grid. The primary input for the design of a turbine is the pneumatic power based upon the pressure amplitude p_{owc} and the

© The Author(s) 2018
A. Moñino et al., *Thermodynamics and Morphodynamics in Wave Energy*,
SpringerBriefs in Energy, https://doi.org/10.1007/978-3-319-90701-7_2

Fig. 2.1 Control volume
scheme. (*Source* [8].
Reproduced with permission
of Elsevier)

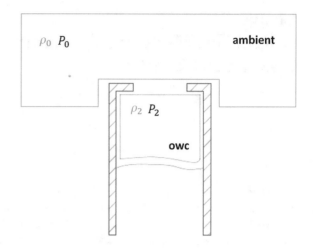

volumetric flow rate Q_{owc} at turbine inlet. The performance indicators are the pressure
drop, power and efficiency, and their dependence on the flow rate. In this section the
basis of the OWC behaviour under the classical formulation is presented.

2.2.1 Classic Formulation

In the classic formulation, the external thermodynamic pressure in the OWC chamber
is set as constant and the pressure inside the chamber is time dependent, according
to Fig. 2.1. The thermodynamic pressures p_0 and p_2 represent the external reference
pressure and the total pressure inside the OWC chamber, respectively. The difference
of these pressures is the manometric pressure inside the chamber (p_{owc}):

$$p_2 = p_{owc} + p_0 . \tag{2.1}$$

Following [7, 11], the continuity equation applied to the control volume of air
inside the chamber reads:

$$Q_T^m = -\frac{d}{dt}(\rho_2 V_{owc}) = -\left(\rho_2 \frac{dV_{owc}}{dt} + V_{owc} \frac{d\rho_2}{dt}\right), \tag{2.2}$$

where Q_T^m is the mass flow through the turbine, ρ_2 is the air density inside the chamber
and V_{owc} is the air volume inside the chamber. It is assumed the process follows the
adiabatic law:

$$pV^\gamma = constant , \tag{2.3}$$

which is a particular case of the general equation $pV^n = constant$ for a polytropic system in a given process. When $n = \gamma = C_p/C_v$ the polytropic equation becomes the adiabatic law described by Eq. 2.3. In general terms, p is pressure and V volume. Consequently

$$\frac{\rho}{\rho_{ref}} = \left(\frac{p}{p_{ref}}\right)^{1/\gamma}, \tag{2.4}$$

taking the solution (ρ_0, p_0) as a reference, the linear relations between pressure and density are established in Eq. (2.1):

$$\rho_2 = \rho_0 + \left(\frac{\rho_0}{\gamma P_0}\right)(p_{owc}). \tag{2.5}$$

In the case of a steady–state flow study in which only the turbine performance is concerned, the mass flow rate conservation simply reads from (2.2):

$$Q^m_{owc} = Q^m_T, \tag{2.6}$$

where the turbine mass flow rate Q^m_T is a function of the pressure drop and the turbine characteristics. Once the mass flow rate Q^m_{owc} from the chamber is fixed, the turbine discharge is fixed and the thermodynamic variables can be observed. A complete review of the OWC classic formulation is shown in Appendix I.

2.2.2 Air–Water Vapour Mixture Density

Some characteristics of the air–water vapour mixture are deduced from the properties of the vapour fraction present in dry air. Using the *Clapeyron-Clausius* law [12]— Fig. 2.2—, the saturation vapour pressure e_s is defined as:

$$e_s = e_0 \cdot exp\left[\frac{L}{R_v}\left(\frac{1}{T_0} - \frac{1}{T}\right)\right], \tag{2.7}$$

where $e_0 = 0.611\,kPa$ is the partial saturation pressure at $T_0 = 273\,K$; L is the latent vaporization heat; $R_v = 461\,J/K \cdot kg$ is the water vapour constant, and $L/R_v = 5423\,K$. The saturation pressure represents the equilibrium state for which no more vapour is present in dry air without further condensation, and it only depends on temperature.

However, the vapour pressure at any temperature—any non–equilibrium state— can be fixed from a given value of relative humidity RH, expressed as the vapour concentration in dry air referred to the vapour concentration at equilibrium:

$$RH = \frac{e}{e_s}, \tag{2.8}$$

Fig. 2.2 Clapeyron-Clausius
relation between saturated
vapour pressure and air
temperature. (*Source* [8].
Reproduced with permission
of Elsevier)

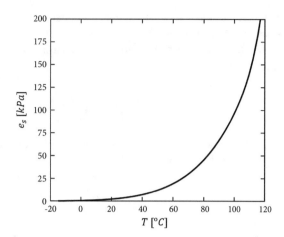

The density of vapour in the air–water vapour mixture, which is know as absolute humidity is r:

$$r = \frac{m_v}{m_a} = \frac{\varepsilon e}{p_g - e} \tag{2.9}$$

where: m_v is the mass of vapour and m_a is the mass of dry air, $\varepsilon = R_a/R_v = 286.7/461 = 0.622$ is the ratio of gas constants for dry air and water vapour respectively; p_g is the pressure of the air–water vapour mixture. To deduce the second equality in (2.9) it must be kept in mind that the pressure of the mixture is the sum of the partial pressures of dry air p_a and water vapour e:

$$p_g = p_a + e \tag{2.10}$$

and that for both dry air and vapour is applied the ideal gas equation.
The density of the gas consisting of moist air can be expressed as de ratio of dry air and vapour masses and the total volume V_g of the gas mixture:

$$\rho_g = \frac{m_a + m_v}{V_g} \tag{2.11}$$

where ρ_g is the real gas density.
The state equations for the partial pressures of dry air and water vapour are:

$$p_a = \frac{m_a}{V_g} R_a T \tag{2.12}$$

and:

$$e = \frac{m_v}{V_g} R_v T \tag{2.13}$$

accounting for the fact that dry air and water vapour occupy the same volume V_g in the mixture. It must be noted from the previous expressions that neither $p_a/R_a \cdot T$ nor $e/R_v \cdot T$ are the densities of dry air and water vapour at the pressure p_g of the mixture.

From (2.9) and (2.11):

$$\rho_g = \frac{m_a(1+r)}{V_g} \tag{2.14}$$

On the other hand, from (2.12) and (2.13):

$$\frac{m_a}{V_g} = \frac{p_a}{R_a T} = \frac{p_g - e}{R_a T} = \frac{p_g}{R_a T} - \frac{1}{R_a T} \frac{m_v R_v T}{V_g}$$
$$= \frac{p_g}{R_a T} - \frac{1}{R_a T} \frac{r m_a R_v T}{V_g} \tag{2.15}$$

Clearing m_a/V_g from both sides and recalling that $\rho_a = p_g/R_a \cdot T$ is the density of dry air at mixture pressure p_g:

$$\frac{m_a}{V_g} = \frac{\rho_a}{1 + r/\varepsilon} \tag{2.16}$$

The density of air–water vapour mixture finally reads from (2.14) and (2.16):

$$\rho_g = \rho_a \frac{1+r}{1+r/\varepsilon} = \rho_a \frac{1+r}{1+1.608r} \tag{2.17}$$

It is clear from (2.17) that the density of moist air is lower than the density of dry air, the amount depending on the humidity through the mixing ratio. Therefore, the conditions of turbine performance can be corrected by means of the water vapour content in dry air. In fact a diminishing in flow density affects the performance of the turbine, in the sense that a less dense air yields a lower drag force on the blades and a lower power generated by the turbine.

2.2.3 Equation of State for the Real Gas

The start point is the equation of state for N moles of an ideal gas:

$$pV = NR_0 T , \tag{2.18}$$

where R_0 is the universal gas constant. The real gas is expected to deviate from the ideal so Eq. 2.18 is not satisfied for all p and T. To deal with the real gas, we start defining the compressibility factor per gas mole, $\mathbb{Z}$, [9], as:

$$\mathbb{Z} = \frac{pV}{R_0 T}, \tag{2.19}$$

For an ideal gas, $\mathbb{Z} = 1$. A modification in the equation of state is carried out using the *Kammerling–Ones equation* (also known as the *Virial* Expansion) [9]:

$$\mathbb{Z} = \frac{pV}{R_0 T} = 1 + \frac{\mathbb{B}}{V} + \frac{\mathbb{C}}{V^2} + \dots, \tag{2.20}$$

where the coefficients $\mathbb{B}$ and $\mathbb{C}$ only depend on the temperature (experimental results).

Hence, an equation of state for a real gas is derived using $\mathbb{Z}$ defined in Eq. (2.20), [13, 14]:

$$p = \mathbb{Z}\rho_g R_g T, \tag{2.21}$$

where ρ_g and R_g are the real gas density and gas constant, determined from Eq. (2.11). Therefore the compressibility factor $\mathbb{Z}$ is expressed from Eq. (2.20) in the alternative form proposed by [13], with truncated series up to the coefficient $\mathbb{B}$:

$$\mathbb{Z} = 1 + \frac{\mathbb{B}p_c}{R_g T_c} \frac{p_r}{T_r}. \tag{2.22}$$

In the previous expression p_c and T_c are the critical values of pressure and temperature for a given gas. Critical temperature and its corresponding pressure value represent the state beyond which a gas cannot be liquefied by compression, and they are standard reference values used for calculations with reduced state variables T_r and p_r, so that $T_r = T/T_c$ and $p_r = p/p_c$. In the case of water $T_c = 647\,\text{K}$ and $p_c = 218\,\text{atm} = 220.89 \times 10^5\,\text{Pa}$—air and water vapour properties are summarised in Appendix II—. The advantage in the previous formulation is that Eq. (2.22) is therefore calculated from [2]:

$$\frac{\mathbb{B}p_c}{R_g T_c} = f_0 + \omega f_1 + \chi_{mol} f_2, \tag{2.23}$$

where: f_0, f_1 and f_2 are temperature correlation functions; ω is the acentric factor; χ_{mol} is the molar fraction of vapour in dry air. In the case of the air-water vapour mixture—hereinafter the real gas—:

$$
\begin{cases}
\omega \simeq 0 \\
f_0 = 0.145 - \dfrac{0.33}{T_r} - \dfrac{0.1385}{T_r^2} - \dfrac{0.0121}{T_r^3} - \dfrac{0.000607}{T_r^8} \\
f_2 = \dfrac{0.0297}{T_r^6} - \dfrac{0.0229}{T_r^8}.
\end{cases}
\tag{2.24}
$$

From the *virial* terms and compressibility factor, it is straightforward to calculate the specific heat, C_{pg}, and the enthalpy, H_g, for the real gas as deviations from the ideal gas values (C_p, H), [14]:

$$
C_{pg} = C_p + \delta C_p p,
\tag{2.25}
$$

and:

$$
H_g = H + \delta H p,
\tag{2.26}
$$

where the deviations of specific heat and enthalpy are:

$$
\delta C_p = -\frac{R_g T_r}{p_c} \frac{d^2}{dT_r^2} (f_0 + \chi_m f_2),
\tag{2.27}
$$

and:

$$
\delta H = \frac{R_g T_c}{p_c} \left[(f_0 + \chi_m f_2) - T_r \frac{d}{dT_r} (f_0 + \chi_m f_2) \right].
\tag{2.28}
$$

2.2.4 Calculation of Thermodynamic Variables

Once the corrected expressions of specific heat and enthalpy have been deduced, the conservation of enthalpy can be applied to a thermodynamic process in a real gas, i.e. the air flow through the turbine in an OWC system. Making use of Eqs. (2.25)–(2.28) in an adapted way from [14], the conservation of enthalpy and flow are applied between turbine inlet and outlet, with known values of local temperature T_{in}, pressure p_{in} and humidity at the inlet. The enthalpy for the ideal gas is expressed as:

$$
H = C_p T + \frac{1}{2} U^2,
\tag{2.29}
$$

where U is the velocity. The velocity at the turbine outlet, U_{out}, is computed from Eqs. (2.26) and (2.29) for the real gas:

$$
U_{out} = \sqrt{2(H_g - C_p T_{out} - \delta H p_{out})},
\tag{2.30}
$$

Here the temperature T_{out} at the turbine outlet should not be evaluated through the adiabatic process equation $T_{out}^{ad} = T_{in}(p_{out}/p_{in})^{(\gamma - 1/\gamma)}$, because strictly speaking this last expression is only valid for ideal gas. Instead of using this last equation, a new value of $T_{out} \neq T_{out}^{ad}$ is guessed until the continuity equation between turbine inlet and outlet is satisfied:

$$\rho_g US\big|_{in} = \rho_g US\big|_{out} . \tag{2.31}$$

The iterative procedure, in which the effect of moisture is implemented through the modified air density, outcomes the new pressure and temperature values for the real gas, required to match the continuity condition. Following the proposed methodology, differences between the ideal gas and the real gas model can be computed and corrections from standard predictions can be applied.

2.3 Experimental Set Up

The turbine performance has been tested with a total of three different environmental conditions: so called "dry air" (non forced humidity, environmental conditions, ≈35% relative humidity), humid air with minimum humidity (50% relative humidity) and humid air with maximum humidity (70% relative humidity).

The wind tunnel is a poly-methyl methacrylate (*PMMA*) structure with a test section that is 3 m in length with a 360 mm × 430 mm cross-section. The wind speed, up to 20 m/s, is controlled by a variable frequency converter controlling an electric fan at the downstream end with a maximum power of 2.2 kW.

The test turbine is mounted on a nylon—casted chamber representing a simplified OWC structure as shown in Fig. 2.6. The turbine has $D_t = 0.03$ m diameter, with cross section area $A_t = 3.5906 \times 10^{-4}$ m^2 (center to blade tip), and turbine blades area $A_a = 2.4271 \times 10^{-4}$ m^2. The turbine solidity is $\sigma = 0.82$, following [10]—a complete description of the turbine performance coefficients as well as the turbine geometry specifications is given in Appendix III—.

The calibration of the turbine reveals a linear response between pressure drop and flow discharge—Fig. 2.3.

The pressure distribution is observed with pressure taps (up to 64 channels per scanner) connected to a scanner working at a sampling period of 0.0016 s. The accuracy of the pressure system is fixed as 0.1% of the measurement range, so the uncertainty in the measures is almost null, as depicted in Fig. 2.3. Six pressure taps are located around the turbine: two of them in the flow direction to measure total pressure, and the rest four samples perpendicular to the flow direction, for the static pressure measurement. There are two static pressure taps and two total pressure taps place upstream the turbine, and two static pressure taps placed downstream.

The air velocity is measured upstream and downstream the turbine with a *Laser Doppler Velocimeter (LDV)*, using as seeding water particles sprayed in the air flow. Taking water drops as seeding particles in air can become a problem: big drops move with a strong delay and generate vertical velocity components; small water particles tend to evaporate, [4]. However, some authors use *LDV* to measure wind speed in tunnel using water drops as seeding, [5]. Locating the water gun outside the

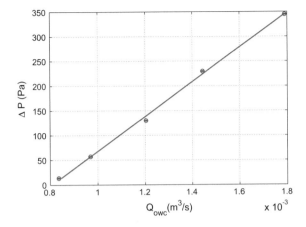

Fig. 2.3 Air flow and pressure drop in turbine. Dry tests. (*Source* [8]. Reproduced with permission of Elsevier)

Fig. 2.4 Wind Tunnel. (*Source* [8]. Reproduced with permission of Elsevier)

wind tunnel with the nozzle pointing toward the honeycomb section ensures that the large water drops are captured by the honeycomb section, [5]. To assure the correct measurement of the *LDV*, a checking measurement with a *Prandtl Tube* was done (Fig. 2.4).

Humidity and temperature are measured with remote sensors located at representative sections *S1* and *S2* upstream the turbine, as shown in Fig. 2.5; the first one is placed near the entrance to control the input humidity. In addition, an external sensor is used to measure the external reference temperature and humidity. Tests are conducted under constant external humidity conditions. The final set up, with the thermo-hygrometer sensors and the LDV measurement points, is shown in Fig. 2.5.

2.4 Results

Experimental results are presented alongside the theoretical calculations according to the methodology for the real gas, as provided in previous sections. From a strictly thermodynamic sense, the state variables defined in the real gas model represent the macroscopic properties of a system under thermodynamic equilibrium, i.e. a state of

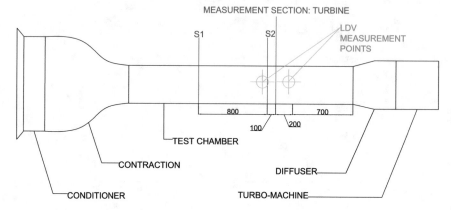

Fig. 2.5 Final set up scheme. (*Source* [8]. Reproduced with permission of Elsevier)

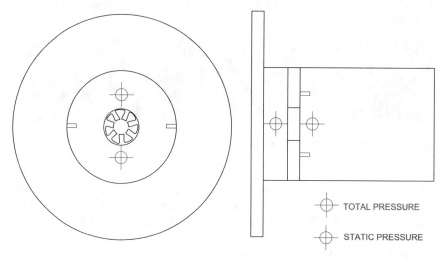

Fig. 2.6 Pressure taps distribution in the measurement section. (*Source* [8]. Reproduced with permission of Elsevier)

an isolated system in which the macroscopic properties are stationary and there is no hysteresis, [12]. Within that scope, a correct understanding of the system response is achieved under a stationary unidirectional flow condition (Fig. 2.6).

2.4.1 Pressure Drop and Air Flow

Pressure drop is plotted as a function of flow discharge in Fig. 2.7 for different humidity values. While for dry air the turbine response is linear according to calibra-

Fig. 2.7 Pressure drop and
air flow in turbine. All tests.
(*Source* [8]. Reproduced
with permission of Elsevier)

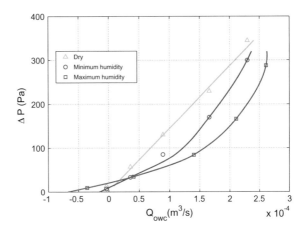

tion, as humidity increases the pressure drop diminishes and changes its functional
dependence for any given discharge. The pressure drop is reduced down to 40%
for discharge values around $2\,\mathrm{m}^3/\mathrm{s}$. Thinking in terms of turbine performance and
efficiency, as moisture increases, greater air flow values are required to preserve a
given pressure drop.

Those results can be generalized in terms of non-dimensional values of pressure
drop and flow discharge, according to Fig. 2.8. In order to account for the humidity
content in the three test cases, the non-dimensional pressure is calculated as:

$$P^* = \frac{p_{owc}}{\rho_2 w^2}\left(\frac{1-RH}{RH}\right), \qquad (2.32)$$

where P^* is the non-dimensional pressure, while the non-dimensional flow discharge is computed from expression (2.41). It can be observed in Fig. 2.8 that non-
dimensional pressure drop can be reduced up to 30% respect to the dry air case for
the highest moisture.

2.4.2 Density Correction by the Air–Vapour Mixture

A first attempt in predicting the turbine response under moist air condition is accomplished by substituting moist air densities calculated by Eq. (2.17), into the classic
continuity Eq. (2.37). Although the values provided by the classic formulation fit
reasonably well with the observed values for dry air, there is no good match for
the turbine response under moist condition, as it is shown in Fig. 2.9. That way, the
classic continuity expression fails in predicting the turbine performance with humid
air, even if the actual turbine also offers a linear response in dry air.

Fig. 2.8 Non-dimensional
pressure drop and air flow in
turbine. All tests. (*Source*
[8]. Reproduced with
permission of Elsevier)

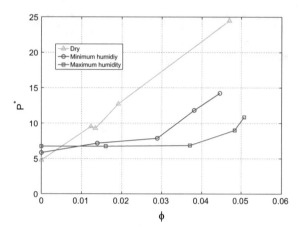

Fig. 2.9 Pressure drop and
air flow. New humidity
curves. Corrected density.
$K = 0.014$. (*Source* [8].
Reproduced with permission
of Elsevier)

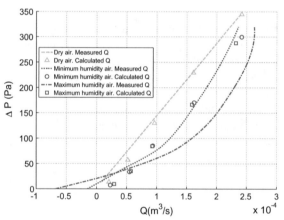

2.4.3   Temperature Calculations for the Real Gas Model and Compressibility Factor $\mathbb{Z}$

Following the methodology proposed for the real gas in previous sections, values of
temperature at the turbine outlet are predicted for the experimental turbine inlet data.
Temperature values are calculated from observation of mass flow and application
of conservation of enthalpy. Differences with respect to the calculated values under
the assumption of adiabatic process for an ideal gas are observed, as it is shown in
Fig. 2.10.

The curves present similar behaviour in all cases. For the dry case, the real gas
model shows a fairly constant temperature value for the complete range of pressure,
and the values are close to those predicted with the ideal gas formulation. However,
the differences between values of temperature for ideal gas prediction and the proposed real gas methodology becomes significant. From the results it is clear that

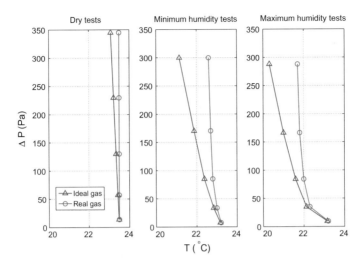

Fig. 2.10 Pressure drop and temperature. All tests. Different temperature calculations. (*Source* [8]. Reproduced with permission of Elsevier)

for the continuity flow condition and conservation of enthalpy to be satisfied in the humid air cases, temperature values must be up to 1 °C above the ideal gas estimation. Thinking in terms of the real gas, the differences can be explained recalling that the adiabatic law, $pV^{\gamma} = constant$, is suitable for an ideal gas, say, a gas addressing the thermal equation of state $p = \rho RT$ for all T. But according to the above dissertation, in a real gas we have $p = Z\rho RT$ which means that ρRT is not always followed exactly. Then the adiabatic law in its standard form might not be fully applicable, and the predictions of temperature

However, the previous results are not incompatible with the hypothesis of adiabatic process. Actually, the adiabatic law $pV^{\gamma} = constant$ is deduced from three fundamental hypothesis: (1) ideal gas equation of state $p = \rho RT$; (2) First Principle of Thermodynamics applied to a thermally isolated system, which means no heat exchange with the exterior; (3) the work exerted on the system takes the form of a reversible process under constant pressure condition. In the case of an air–water vapour mixture, the hypothesis of thermal isolation and reversible process are applicable as in other local atmospheric processes. But the presence of vapour in air and the subsequent induced changes as a function of temperature—mixture density, saturation pressure, etc.—induce a deviation from the ideal gas law. In fact, it is assumed that the flow through the turbine is adiabatic, but with a new law different from the ideal one. This can be further observed in the calculations of the compressibility factor $\mathbb{Z}$.

The compressibility factor $\mathbb{Z}$ is calculated following Eq. (2.22) and plotted in Fig. 2.11 for the case study. The compressibility factor gives information about the deviation of the real gas behaviour from the expected ideal response. The behaviour of $\mathbb{Z}$ is linear for the whole range studied, i.e. between 22.6 and 23.2 °C, with overlapping

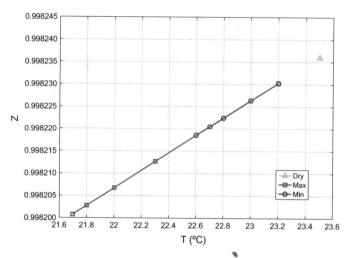

Fig. 2.11 Compressibility factor $\mathbb{Z}$. All tests. (*Source* [8]. Reproduced with permission of Elsevier)

Fig. 2.12 Compressibility factor $\mathbb{Z}$. (*Source* [8]. Reproduced with permission of Elsevier)

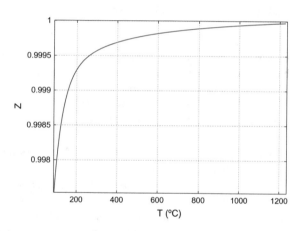

of minimum and maximum humidity curves. The values keep away from 1 (that value would be equivalent to an ideal gas) as temperature decreases.

However, the linear trend in $\mathbb{Z}$ is only preserved for low temperature values, as shown in Fig. 2.12, while for high temperatures—say for reduced temperatures $T_r \gtrsim 0.7$—tends to an asymptote $\mathbb{Z} \to 1$, accordingly to the real gas model.

As temperature decreases the gas behaviour differs from the ideal one. This result is consistent with others obtained: humidity in air makes the air–water vapour mixture different from an ideal gas.

2.4.4 Power Input

Non-dimensional power input calculated from non-dimensional values of pressure drop and flow discharge, following Eq. (2.33). Results are plotted in Fig. 2.13.

$$W^* = P^*\Phi .$$ (2.33)

Following the same argument as in previous results, the power input to the turbine can be reduced up to 33% respect to the dry air value as moisture increases. The effect of moisture on turbine power input can be further observed when the power input ratio values referred to the lowest moisture content and plotted against the flow coefficient, as depicted in Fig. 2.14. Minimum values of the power input ratio ranging from 0.5 to 0.3 are found for when RH exceeds 0.55, depending on the flow coefficient. An asymptotic value of 0.4 is reached for $RH \geqslant 0.65$, indicating the moisture has no added effect on turbine performance beyond that point. It must be recalled the results correspond to a stationary flow condition for a generic linear turbine (not a Wells type turbine). It can be inferred that a reduction in the maximum power input can be expected for standard OWC performance during transient cycles of compression and expansion. The extent of such reduction has to be determined for a given moisture during a representative compression–expansion cycle, according to the time–dependent values of the flow coefficient.

Within the scope of this chapter, the reduction in the expected power could be significant when applied to the long–term management of an operative prototype. The greater the moisture content in the standard working conditions of a energy extraction device, the greater the reduction in the whole performance. Although the amount of that reduction should be studied for each particular design, the results reveal the effect of moisture on power input and subsequent performance might not be neglected.

Fig. 2.13 Non-dimensional power input and air flow. All tests. (*Source* [8]. Reproduced with permission of Elsevier)

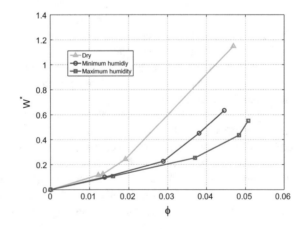

Fig. 2.14 Power input ratio
and moisture effect. (*Source*
[8]. Reproduced with
permission of Elsevier)

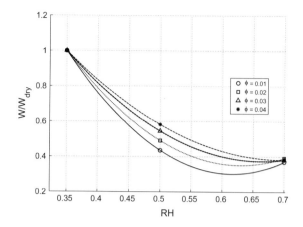

2.5   Discussion and Future Research

The final goal of this research is to apply a real gas approach to calculate the power performance of a system OWC. The work takes the line left open by [14]. They applied the model and left a proposed line of experimentation: this chapter applies the experimental and check the results with the model. As a first step in this research the tests were performed with one-way flow. The turbine used in the experiments is not reversible. However, the linear turbine guarantees the similarity with a Wells turbine. Therefore, in addition, it is not necessary to consider transient states due to oscillatory turbine flow and no further restrictions on the turbine type or turbine dynamics are required.

Although this is a generalization of the OWC turbine problem, the basic thermodynamics of the real gas can be observed in a simplified manner without loss of applicability to the case of an oscillatory flow. Despite this, for further work and a more complete study of this phenomena, Wells turbine tests will be necessary in the future. Indeed the next step will be conducting tests with a reversible turbine in oscillatory flow to analyse harmonic.

On the other hand, it is highly recommendable to observe real values of efficiency operating plants, and to check if there are evidences of turbine efficiency reduction in the real OWC plants.

The hypothesis of incompressible flow is present throughout the work, addressing the same focusing applied by other authors, [6]. Actually no further considerations of compressibility are assumed, since values of Mach number $M = U_0/C_s$—defined as the ratio between a representative flow velocity and the speed of sound in air—are no greater than 0.003. Those values are far below the limit of 0.3 and the flow can be assumed to be incompressible. For $M \gtrsim 0.3$ the flow starts to deviate form the incompressible response, and the real observed pressure values might be 2%

greater than those predicted by incompressibility calculations, [1]. On the other hand, compressibility effects might be present when conducting tests at different scales and when extending the results to full–scale prototypes, due to velocity scales leading to increasing Mach numbers. To predict the scale effects and the deviations between incompressible expected values and observed variables, the theoretical model has to account for the effect of compressibility. The connection between thermodynamic variables—pressure, temperature, density—and flow velocity can be fixed through the state equation of enthalpy, [3].

2.6 Conclusions

The purpose of the present chapter is to study the thermodynamic of a real gas flow through the turbine and its differences with respect to the ideal gas hypothesis, with the final goal to be applied to OWC systems. The application of a real gas model and the experimental study confirmed the deviations of the turbine performance from the expected values depending on flow rate, moisture and temperature. The main conclusions of this research are:

1. For a turbine with linear performance in terms of pressure drop and air flow rate under dry air conditions, the experiments reveal that the turbine behaviour under stationary-flow condition deviates from linearity for the air-water vapour mixture, with differences in pressure drop up to 40% from the dry air case.
2. Following the proposed real gas methodology, it is found that the conservation of air flow rate and enthalpy requires a temperature variation between turbine inlet and outlet that does not follow the standard adiabatic law for the ideal gas. In fact, the compressibility factor shows values that deviates from 1, which means the real gas differs slightly from the ideal model. Hence the state equation for the real gas differs from the ideal case, and therefore its behaviour keeps away from the ideal adiabatic description of the process.
3. The power input, understood as the available power for the turbine, decreases when moisture increases. The non-dimensional power input ratio referred to the lowest moist content state, shows a diminishing between 50 and 70% in the expected power input with respect to the dry air condition values, with an asymptotic response as moisture content exceeds 65%.
4. Even though those values have been obtained for a generic linear turbine in stationary regime, the results indicate that such a deviation should be expected for standard OWC devices equipped with Wells turbines.

Appendix I: OWC Classic Formulation

In order to appropriately set the context and the scope of the study, a brief reminder on the classic OWC formulation is presented. If a Wells turbine—or any linear response turbine—is used for the power take-off system, the mass flow rate Q_T^m through the turbine can be defined in terms of the pressure drop in linear form:

$$Q_T^m = \rho_0 Q_T = \rho_0 \frac{KD_t}{N\rho_0} p_{owc} , \qquad (2.34)$$

where K depends on turbine parameters, following [7], D_t is the turbine diameter and N is the turbine rotation velocity.

Substituting into Eq. 2.2, multiplying by ρ_0 and clearing up Q_{owc}:

$$Q_{owc} = \frac{KD_t}{N\rho_2} p_{owc} + \frac{V_{owc}}{C_s^2 \rho_2} \left(\frac{dp_{owc}}{dt} \right) , \qquad (2.35)$$

where C_s is the speed of sound in air.

Following [7] and applying harmonic solutions, we obtain:

$$\widehat{Q}_{owc} = \frac{KD_t}{N\rho_2} \widehat{p}_{owc} - i\omega \frac{V_{owc}}{C_s^2 \rho_2} \left(\widehat{p}_{owc} \right) , \qquad (2.36)$$

where both $\widehat{Q}_{owc}$ and $\widehat{p}_{owc}$ are complex amplitudes defined in the classic problem.

If as discussed before a steady–state flow is considered, the continuity equation simply reads:

$$Q_{owc} = \frac{KD_t}{N\rho_2} p_{owc} , \qquad (2.37)$$

with Q_{owc} and p_{owc} defined as real quantities. In general terms, the power input to the turbine is expressed as:

$$W_I = p_{owc} Q_{owc} . \qquad (2.38)$$

where W_I is the power input.

Following [7], the average power $\overline{W}$ extracted from the pneumatic power inside the chamber is:

$$\overline{W} = \frac{1}{2} \frac{KD_t}{N\rho_0} |\widehat{p}_{owc}|^2 = \frac{1}{2} \frac{KD_t}{N\rho_0} (\rho_w g)^2 \frac{|\tilde{\Gamma}'|^2 (H/2)^2}{(\chi + \tilde{B})^2 + (\tilde{C} + \beta)^2} , \qquad (2.39)$$

and the capture length, understood as an efficiency expression, is:

$$kL = \frac{8k\overline{W}}{\rho_w g H^2 C_g} = \frac{gkR_{owc}}{\omega C_g} \frac{\chi |\tilde{\Gamma}'|^2}{(\chi + \tilde{B})^2 + (\tilde{C} + \beta)^2} . \tag{2.40}$$

where kL is the capture length. Coefficients $\tilde{B}$, $\tilde{C}$, $\tilde{\Gamma}$, χ and β are the mathematical representation of the oscillation damping and the restitution of the added mass, the flow in the chamber associated to the diffraction problem and the turbine characteristics. H is the wave height, C_g is the wave group celerity, $g = 9.81 \text{ m/s}^2$ is the gravity acceleration, k is the wave number and R_{owc} is the OWC radius.

Appendix II: Air and Water Vapour Properties

See Table 2.1.

Table 2.1 Dry air and water vapour properties

	Value	Units
Air properties		
R_a	286.7	J/kg · K
$C_{p,a}$	1010	J/kg · K
ρ_a	1.25	kg/m^3
MW_a	0.0288	kg/mole
$T_{c,a}$	132	K
$p_{c,a}$	37.71×10^5	Pa
Water vapour properties		
R_v	461	J/kg · K
$C_{p,v}$	1093	J/kg · K
MW_v	0.0182	kg/mole
$T_{c,v}$	647	K
$p_{c,v}$	220.89×10^5	Pa

Appendix III: Turbine Performance Coefficients

In general terms, the turbine performance is described by the following non-dimensional coefficients:

- Flow coefficient:

$$\Phi = \frac{V_x}{U},\qquad\qquad(2.41)$$

- Pressure drop:

$$P^* = \frac{p_{owc}}{\rho_2 w^2},\qquad\qquad(2.42)$$

- Solidity:

$$\sigma = \frac{A_a}{\pi(D_t/2)^2},\qquad\qquad(2.43)$$

where V_x is the axial velocity, $U = ND_t/2$ is the blade circumferential velocity, w is the relative incoming velocity, A_a is the total blades area.

In the following Table 2.2, the turbine geometry specifications are given.

Table 2.2 Turbine geometry specifications

Property	Value
R_{int}	0.0055 m
R_{ext}	0.015 m
R_{med}	0.0102 m
D_t	0.03 m
A_t	3.5906×10^{-4} m^2
c	0.0076 m
Number of blades	7
A_a	2.4271×10^{-4} m^2
σ	0.82

References

1. Blevins, R.D.: Applied fluid dynamics handbook. In: Van Nostrand Reinhold Company. Library of Congress Catalog Card Number, pp. 83–14517 (1997)
2. Evans, D.V.: Wave power absorption by systems of oscillating pressure distributions. J. Fluid Mech. **114**, 481–499 (1982)
3. Johnson, R.W.: The Handbook of Fluid Dynamics (1998)
4. Longo, S., Losada, M.A.: Turbulent structure of air flow over wind'induced gravity waves. Exp. Fluids **2**(53), 369–390 (2012). https://doi.org/10.1007/s00348-012-1294-4
5. Longo, S.: Wind-generated water waves in a wind tunnel: Free surface statistics, wind friction and mean air flow properties. Coast. Eng. **1**(61), 27–41 (2012). https://doi.org/10.1016/j.coastaleng.2011.11.008
6. López, I., Castro, A., Iglesias, G.: Hydrodynamic performance of an oscillating water column wave energy converter by means of particle imaging velocimetry. Energy **83**, 89–103 (2015)
7. Martins-Rivas, H., Mei, C.C.: Wave power extraction from an oscillating water column at the tip of a breakwater. J. Fluid Mech. **626**, 395–414 (2009)
8. Medina-López, E., Moñino, A., Clavero, M., Del Pino, C., Losada, M.A.: Note on a real gas model for OWC performance. Renew. Energy **85**, 588–597 (2016)
9. Prausnitz, J., Lichtenthaler, R., Gomes de Azevedo, E.: Molecular Thermodynamics of Fluid–Phase Equilibria. Prentice–Hall (1999). ISBN 0-13-977745-8
10. Raghunathan, S.: The wells turbine for wave energy conversion. Prog. Aerosp. Sci. **31**, 335–386 (1995)
11. Sarmento, A., Gato, L., de O Falcão, A.F.: Turbine-controlled wave energy absorption by oscillating water column devices. Ocean Eng. **17**(5), 481–497 (1990)
12. Stull, R.B.: Meteorology for Scientists and Engineers, Pacific Grove, CA (2000)
13. Tsonopoulos, C., Heidman, J.L.: From the virial to the cubic equation of state. Fluid Phase Equil. **57**, 261–276 (1990)
14. Yang, W., Su, M.: Influence of moist combustion gas on performance of a sub-critical turbine. Energy Convers. Manage. **46**(1), 821–832 (2004)

Chapter 3
Thermodynamics of an Oscillating Water Column Containing Real Gas

Abstract Oscillating Water Column (OWC) devices are usually modelled as simple systems containing ideal, dry air. However, high humidity levels are likely to occur in a prototype device open to the sea, particularly in warm climates such as prevail in the lower latitudes. In this chapter, a real gas model is implemented to take into account humidity variations inside an OWC chamber. Using a modified adiabatic index, theoretical expressions are derived for the thermodynamic state variables including enthalpy, entropy and specific heat. The model is validated against experimental data, and shown to provide better agreement than obtained using the ideal gas assumption. By calculating real air flow in an OWC it is shown that the mechanical efficiency reduces and the flow phase alters with respect to the ideal gas case. Accurate prediction of efficiency is essential for the optimal design and management of OWC wave energy converters.

3.1 Objective

The aim of this chapter is to devise a mathematical formulation for the polytropic process of a real gas that describes the adiabatic compression-expansion cycle of a dry air–water vapour mixture in an OWC wave energy converter. The real gas formulation allows us to determine the deviation in PTO efficiency from the expected value under an ideal gas assumption. Experimental data from wind tunnel testing of an OWC chamber will be used to validate the model by comparing values of state variables against predictions by the deduced process equation.

3.2 Methodology

3.2.1 Adiabatic Process of a Real Gas: General Approach

We now derive the equation for a polytropic process in a real gas system. Figure 3.1 illustrates an OWC system from a thermodynamic perspective. The system comprises

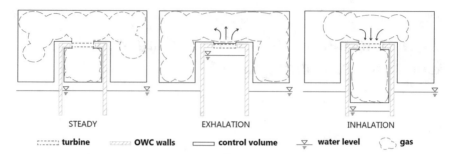

STEADY   EXHALATION   INHALATION

┈┈┈ turbine   ┄┄┄ OWC walls   ▭ control volume   ▽ water level   ◯ gas

Fig. 3.1 Definition sketch of OWC general scheme, showing control volume. (*Source* [6]. Reproduced with permission of Elsevier)

a chamber, turbine, and external surrounding air. The control volume is independent of the nature of the gas inside the system.

The model is based in the following hypothesis:

- The system is closed. Exchanges of mass and heat in the OWC system take place between the chamber and surrounding air in the vicinity of the device. Provided the surrounding region is sufficiently large, then mass exchange between that region and the atmosphere can be neglected without loss of generality.
- The system is isentropic. It is assumed that heat exchange does not occur between the OWC system and either the atmosphere or the surrounding water. This is not a hard restriction. In terms of the exchange process during a compression-expansion cycle, the heat required to address an isothermal process would be greater than that eventually exchanged during the cycle, [4], leading to an essentially adiabatic process.
- All processes are reversible. If the OWC system is considered similar to gas forced by a frictionless piston (i.e. the free surface inside the chamber) it may be assumed that the OWC system can be returned to the previous state by simply relaxing the compression force for any compression state inside the chamber. Otherwise, the deformation work δL exerted on the air chamber cannot be expressed as $-p\delta V$.

In consequence, the starting conditions from which to deduce the equation of a polytropic process of a real gas are the same as those for an ideal gas. The only distinction lies in the equation of state for the real gas. Here, the real gas is described by either the Virial equation of state, or the Kammerling–Onnes expansion, [1, 13], which in its Leiden form is given by [9], as

$$\frac{pv}{R_0 T} = 1 + \frac{B}{v} + \cdots \tag{3.1}$$

where p is pressure, $v = V/N$ is the molar volume, with V the volume and N representing the number of moles, $R_0 = 8.31\,\mathrm{J/mole \cdot K}$ is the universal gas constant, B is the second virial coefficient (to be determined later), and T is temperature. The real gas equation is defined by analogy to the ideal gas equation, [1], as

$$pv = ZR_0T, \tag{3.2}$$

where Z is the compressibility factor, which models the difference between the real gas and equivalent ideal gas.

3.2.2 Polytropic System and Adiabatic Index n for a Real Gas

A thermodynamic process is a succession of different states, progressively altering from an initial equilibrium state to a final one. In this case, the process comprises a compression/exhaust cycle within a polytropic system. The following derivations are based on the standard polytropic formulation (see e.g. [1, 8], or [2]). The polytropic equation for a gas process is:

$$pv^n = constant, \tag{3.3}$$

where n is the polytropic index. The *General Process Equation* (3.4), [8], expresses the relationship between p and v for a general process in a closed system, in which the variable y is constant without additional restrictions, as

$$\left(\frac{\partial p}{\partial v}\right)_y = -\frac{m}{vk_T}, \tag{3.4}$$

where k_T is the isothermal compressibility factor, and m is an index defined in terms of the specific heats as:

$$m = \frac{C_y - C_p}{C_y - C_v}. \tag{3.5}$$

Specific heats are defined, in general terms, as $C_y = T\left(\partial s/\partial T\right)_y$, where s is the molar entropy, and y depends on the nature of the process. For an isothermal process $y = T$, for an isobaric process $y = p$, and for an adiabatic process $y = s$. Taking the polytropic Eq. (3.3) in its differential form, together with the *General Process Equation* (3.4) and operating, a general expression for the adiabatic index is obtained:

$$n = \frac{m}{pk_T}, \tag{3.6}$$

where the compressibility factor can be expressed as $k_T = -\frac{1}{v}\left(\frac{\partial v}{\partial p}\right)_T$. This expression for n is generally applicable to any process in a closed system. Note that if the process is adiabatic for an ideal gas, then $k_T = 1/p$, with Eq. (3.6) reducing to $n = C_p/C_v$. In this case, the polytropic index takes a constant value of $n = 1.4$ for diatomic molecules, and $n = 1.67$ for monoatomic molecules.

Next, consider a real gas. Here the real gas equation is obtained as a modification of the ideal gas Eq. (3.2). Substituting the partial derivative $\left(\frac{\partial v}{\partial p}\right)_T$ into the general expression of polytropic index (3.6), and eliminating the isothermal compressibility factor k_T, noting the definitions of the specific heats C_p and C_v, the resulting general expression for n is

$$n = \frac{m}{1 - \frac{p}{Z}\left(\frac{\partial Z}{\partial p}\right)_T}. \tag{3.7}$$

For an **adiabatic** process, no heat is transferred to the surrounding universe, and Eq. (3.7) then becomes

$$n = \frac{C_p/C_v}{1 - \frac{p}{Z}\left(\frac{\partial Z}{\partial p}\right)_T} \tag{3.8}$$

Equation (3.8) allows us to represent mathematically a real gas process in the continuity equation, applied to an OWC air chamber, given that $pv^n = const$. For the hypothetical case of an ideal gas, then $Z = 1$ and Eq. (3.8) reduces to $n = C_p/C_v$, the usual form. The specific heat coefficients, C_p and C_v, refer to a real gas, with expressions that relate to the virial coefficients through thermodynamic formalism. Consequently, Eq. (3.8) represents the polytropic index for a general process in a real–gas system.

A simpler form of n can be obtained using the definition of the compressibility factor Z given by [12],

$$Z = 1 + \frac{B' p_c p_r}{R_g T_c T_r} = 1 + \frac{B p_c p_r}{R_0 T_c T_r}, \tag{3.9}$$

where $B' = B/M$ with M the molar weight. Note that $p_r = \frac{p}{p_c}$ is the reduced pressure, and $T_r = \frac{T}{T_c}$ is the reduced temperature. Combining (3.8) and (3.9), the following simplified expression is obtained for the adiabatic index n for a real gas:

$$n = Z\frac{C_p}{C_v} \tag{3.10}$$

The compressibility factor is deduced from the Tsonopoulos-Heidman innovation, and the specific heats for the real gas are expressed as functions of ideal gas values, as (defined in the next section). This form of n is simple, compact and easy to work with.

Figure 3.2 presents a contour plot of the functional dependence of n on reduced pressure p_r and reduced temperature T_r, expressed by Eq. (3.10). Here, the

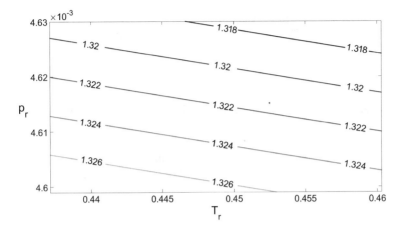

Fig. 3.2 Contour plot of the variation in adiabatic index n with reduced temperature and reduced pressure for a real gas. (*Source* [6]. Reproduced with permission of Elsevier)

Table 3.1 Value of the process equation constant obtained using ideal and real gas model

Constant value for $pv^n = constant$

(T_r, p_r)	Ideal	Real
$(0.44, 4.605 \times 10^{-3})$	7.804×10^4	7.851×10^4
$(0.45, 4.620 \times 10^{-3})$	7.804×10^4	7.851×10^4
$(0.46, 4.635 \times 10^{-3})$	7.804×10^4	7.851×10^4

temperature range is from 10 to 25° C and the pressure from 0 to 2500 Pa. There is a discernible linear dependence of n on pressure. Taking a reduced pressure $p_r = 4.59 \times 10^{-3}$, equivalent to the atmospheric pressure, the adiabatic index n reaches a maximum at 1.33, decreasing linearly towards 1.3 as the relative pressure increases.

Next, consider the constant in Eq. (3.3), which is assumed independent of the modified values of v and p for a real gas, and therefore should be insensitive to n. Table 3.1 lists values of the constant for different combinations of relative temperature and pressure. This confirms the initial hypothesis that the adiabatic process remains adiabatic, independent of the value of n.

3.2.3 Speed of Sound in a Real Gas

The speed of sound, C_s, features in the analysis of wave energy extraction devices through the relationship between pressure and density, and affects compressibility. Although the speed of sound is not utilised directly herein, it should be noted that C_s

must be modified following any change to n in future calculations related to OWC chambers. The speed of sound in an ideal gas undergoing an adiabatic process is [2],

$$C_s^{*2} = \left(\frac{\partial p}{\partial \rho}\right)_0 = \frac{np_0}{\rho_0}.$$

(3.11)

For an adiabatic process involving a real gas, Eq. (3.11) can be modified by applying the real gas equation, $p = \rho Z R_0 T$, utilizing the definition of the adiabatic index for a real gas. Following the same approach taken previously to n, the expression of the speed of sound in a real gas can be simplified using Eq. (3.9) to give

$$C_s = \sqrt{\frac{C_p}{C_v} Z^2 R_0 T}.$$

(3.12)

3.2.4 Specific Heats (C_p and C_v), Entropy (s), Internal Energy (u), Enthalpy (h) and Chemical Potential (μ) for a Real Gas

State variables play a major part in the description of a real gas process. To build up a complete framework of the thermodynamic behaviour of an OWC, expressions for the real gas state variables may be conveniently derived from the ideal gas model. From now on, the ideal gas magnitudes are identified by the superscript index "$*$". Starting from the definition of C_p, and applying the Maxwell relations to the definition of the molar entropy s, the differential form of s can be obtained. Applying then the virial expansion, [3], and operating, the following expression for the molar entropy of a real gas is obtained as [1],

$$s(T, p) = s^*(T, p) - \frac{dB}{dT}p.$$

(3.13)

Using the real gas molar entropy, the real gas specific heat at constant pressure is given by:

$$C_p \approx C_p^* - T\frac{d^2 B}{dT^2}p.$$

(3.14)

For C_v the approximation is not straightforward. Using the differential form of the molar internal energy u, and integrating between two pressure states (p_0 and p), the following expression for the real gas internal energy is obtained:

$$u(T, p) = u^*(T) - T^2\frac{R_0 dB}{v\ dT}.$$

(3.15)

Inserting Eq. (3.15) into the definition of C_v gives

$$C_v \approx C_v^* - \frac{R_0}{v} \frac{d}{dT} \left(T^2 \frac{dB}{dT} \right),$$ (3.16)

where B is the second virial coefficient, defined by [12] as

$$\frac{Bp_c}{R_0 T_c} = f_0 + \omega f_1 + \chi_{mol} f_2.$$ (3.17)

The f coefficients, also called *temperature correlation functions*, are calculated through the Tsonopoulos-Heidman approximation [12], which expresses the coefficients as functions of the reduced temperature of the real gas, T_r, given by:

$$\begin{cases} f_0 = 0.145 - \dfrac{0.33}{T_r} - \dfrac{0.1385}{T_r^2} - \dfrac{0.0121}{T_r^3} - \dfrac{0.000607}{T_r^8}, \\[2mm] f_1 = 0.0637 + \dfrac{0.331}{T_r^2} - \dfrac{0.423}{T_r^3} - \dfrac{0.008}{T_r^8}, \\[2mm] f_2 = \dfrac{0.0297}{T_r^6} - \dfrac{0.0229}{T_r^8}. \end{cases}$$ (3.18)

where ω is the accentric factor, whose value is almost zero ($\omega \simeq 0$) for symmetric molecules such as H_2O. In Eq. (3.17), χ_{mol} is the molar fraction of water vapour in dry air for a given real gas.

In a similar way as for the specific heats, the following expressions for enthalpy and chemical potential of a real gas are obtained:

$$h(T, p) = h^*(T, p) + Bp - T \frac{dB}{dT} p,$$ (3.19)

and

$$\mu(T, p) = \mu^*(T, p) + Bp.$$ (3.20)

The ideal gas values are summarised in Appendix I.

3.2.5 Non Dimensional Thermodynamic Parameters for a Real Gas

The thermodynamic parameters derived in Sect. 3.2.4 are non–dimensionalized in order to gain a universal perspective of their behaviour under different moisture levels. Using the *Buckingham Π Theorem*, we obtain the following non dimensional numbers:

$$\Delta \tilde{s} = |\frac{T^3 R_0}{p^2 Bv}| \Delta s, \tag{3.21}$$

$$\tilde{C}_p = |\frac{T}{pB}| C_p, \tag{3.22}$$

$$\tilde{u} = |\frac{T R_0}{p^2 Bv}| u, \tag{3.23}$$

$$\tilde{h} = |\frac{T R_0}{p^2 Bv}| h. \tag{3.24}$$

Figures 3.3a, b shows the variations in $\Delta \tilde{s}$, $\tilde{C}_p$, $\tilde{u}$ and $\tilde{h}$ with T_r for three relative humidity conditions, dry air $(RH = 0\%)$, air of medium humidity $(RH = 50\%)$, and saturated air $(RH = 100\%)$. All variables are expressed per mole unit. The reduced temperature range corresponds to that of the experiments conducted by [7].

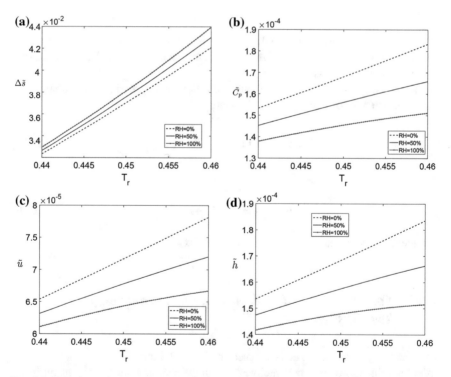

Fig. 3.3 Variations of **a** non dimensional entropy change, **b** specific heat at constant pressure, **c** internal energy, and **d** enthalpy with reduced temperature for a real gas. (*Source* [6]. Reproduced with permission of Elsevier)

Figure 3.3a indicates that entropy progressively increases with temperature, due to an associated increase in heat in the real gas system. Moreover, saturated gas is more entropic than dry gas, which implies a lower level of energy is available for the saturated gas. This behaviour is corroborated by comparing the entropy–temperature curve with the enthalpy and internal energy counterparts. Energy for saturated gas is lower than for dry gas. And dry conditions are closer to ideal gas. Thus it can be concluded that real gas is more entropic than ideal gas, which translates to lower available energy for real gas. Figure 3.3c shows that the specific heat at constant pressure is consistently lower for a real gas. The heat needed to increase the temperature of a real gas by one degree is lower than for an ideal gas because of intermolecular forces which are only represented by the real gas formulation. In the case of a real gas, when heat is supplied to the system, molecular agitation increases because of repulsive intermolecular forces that promote collisions. This raises the kinetic energy of particles in the system, and so increases the temperature, resulting in less external energy available to increase the system temperature. This leads to lower specific heat for a real gas than for an ideal gas.

The finding can be further explained by considering a scenario where the ideal gas represents a mix of oxygen and hydrogen, whose specific heats at constant pressure are $C_p(O_2) = 0.9\,\text{J/K} \cdot \text{g}$ and $C_p(H_2) = 14.3\,\text{J/K} \cdot \text{g}$. If there is twice the amount of hydrogen than oxygen, then the specific heat of the mixture will be 9.8 J/K ·g. So, if an equivalent real gas is composed of a molecule of water vapour (formed by independent hydrogen and oxygen molecules present in the ideal gas), the specific heat of the real gas will now be 2 J/K · g.

3.2.5.1 The Real Gas Non-dimensional Number

A noticeable result is the appearance of a common non–dimensional group, $T R_0 = p B$, in Eqs. (3.21) to (3.24). The group appears as a primary factor in the real gas Eq. (3.9). In general terms, the group is defined as

$$Rg = -\frac{T R_0}{p B} = \frac{1}{1 - Z}. \tag{3.25}$$

Note that Rg takes the following limits depending on the nature of the gas:

$$\begin{cases} Rg \to \infty, \; as \; Z \to 1 \; (ideal gas), \\ Rg \to 1, \; as \; Z \to 0. \end{cases} \tag{3.26}$$

The closer Rg is to 1, the closer the system is to real–gas behaviour. Although the value of $Z = 0$ is hypothetical, it provides a useful means by which to obtain a limit that can be used to compare real and ideal gas behaviours. For example, a mixture of dry air and water vapour with a pressure distribution of $[-100, 100]$ kPa, relative humidity between [0–100%] and temperature range [15, 20]°C, has values of $Z \in [0.99 - 0.998]$ and related values of $Rg \in [80 - 500]$.

3.3 Experimental Validation

To validate the proposed thermodynamic model, predictions are compared against experimental data obtained by [7] on steady air-water vapour mixture flow through a chamber and turbine. Here, the real gas model is used to predict the change between inner and outer system variables, while preserving mass flow rate continuity through the turbine. It should be noted that conservation of mass was not ensured when applying the adiabatic equation for an ideal gas to calculate the temperature at the physical OWC outlet used in the laboratory tests. Herein experimental data are inserted in the formulation of enthalpy to calculate the theoretical outlet velocity, and then used to obtain the mass flow balance in the OWC chamber. Figure 3.1 indicates the key stages involved in the laboratory tests performed by [7]. Experimental data were acquired on flow velocity, air temperature, and pressure at the turbine inlet (U_{in}, T_g, p_g), relative humidity in the chamber RH, and pressure at the turbine outlet p_{out}. During the tests, measurements were made of RH at the outlet and U_{out}, but due to non-correspondence between the ideal law applied to the theoretical control volume and the measured variables, a real gas calculation proved necessary in order to confirm the starting hypothesis (real gas under adiabatic process). Commencing from the modified adiabatic index for a real gas, Eq. (3.10), values of Z, C_p and C_v for a real gas are input from the experimental data. Here, the adiabatic index for real gas is obtained from Eq. (3.10) as a function of Z, C_p, C_v, and the outlet adiabatic temperature from the real gas calculations. Moreover, a real gas calculation for the outlet velocity is needed. Finally, the real gas density is calculated.

Following [7], the procedure is summarised in Appendix II. Comparison is undertaken between the outlet mass flows obtained by applying the ideal and real gas models to the adiabatic process at the turbine outlet, with respect to the inlet mass flow. At steady-state equilibrium, the inflow and outflow mass flow rates are equal. Table 3.2 lists the measured inlet and theoretical outlet mass flow rates. There is very good agreement between the predicted mass flow rate obtained using the real gas formulation and the measured mass flow rate. This is not the case for the ideal gas predictions. The present calculations, which use the new approximation for the adiabatic index n for real gas, ensure mass conservation holds within the system.

The ideal gas formulation does not fully guarantee conservation, and so should not be applied when humidity is present. The modified definition leads to a revised adiabatic index that is below the conventional value for dry air of 1.4. Moreover, the fact that mass conservation is achieved through the real gas formulation confirms the starting hypothesis. Here, the real gas hypothesis enables the theoretical model to represent properly actual conditions present in OWC wave energy converters.

Table 3.2 Mass flow conservation tests, comparing measured inflow rates against predicted outflow rate using ideal and real gas models

Mass flow rate (kg/m² s)		
In	Out adiabatic ideal	Out adiabatic real
Dry tests		
2.077	6.198	2.054
5.667	13.296	5.605
9.174	20.303	9.074
12.559	27.109	12.423
15.738	33.397	15.567
Minimum humidity tests		
2.242	4.848	2.218
3.426	9.790	3.390
7.989	16.654	7.905
11.146	23.469	11.030
14.296	30.920	14.148
Maximum humidity tests		
2.502	5.488	2.475
5.437	10.825	5.381
7.683	16.432	7.607
9.926	22.672	9.828
12.543	29.614	12.421

3.4 Application to OWC Formulation

Using a similar methodology to that of [10], who considered an ideal gas, the thermodynamic theory developed herein for real gas is applied to the basic OWC chamber. The subscript "g" is assigned to variables inside the chamber. The mass of air inside the chamber is

$$m = \rho_g V, \tag{3.27}$$

where the density ρ_g is that of a real gas. The flow rate driven by the water surface movement is defined as

$$Q_w = -\frac{dV}{dt}. \tag{3.28}$$

Exhalation occurs when $dm/dt < 0$, and inhalation when $dm/dt > 0$. This implies that the chamber pressure is greater than atmospheric during exhalation, and the air is decompressed during inhalation. During exhalation, pressurized air is driven out through the PTO system, whereas during inhalation atmospheric air is sucked through the PTO. Hence,

$$\begin{cases} Q_p = -\frac{1}{\rho_0} \frac{dm}{dt}, & \text{inhalation,} \\ Q_p = -\frac{1}{\rho_g} \frac{dm}{dt}, & \text{exhalation.} \end{cases} \tag{3.29}$$

In both cases, the power in the chamber is calculated as $P_w = pQ_w$, taking into account the difference between exhalation and inhalation. The power available to the PTO is $P_{PTO} = pQ_p$.

For an adiabatic process involving a real gas, the relationship between density and pressure is given by

$$\frac{p}{\rho_g^n} = constant. \tag{3.30}$$

Then, linearising:

$$\rho_g = \rho_0 \left(1 + \frac{p}{np_0} \right). \tag{3.31}$$

Note that n, the adiabatic index for a real gas, depends on temperature and pressure changes, which vary in time. Differentiating Eq. (3.31) and substituting the result into (3.29) gives,

• Inhalation

$$Q_p = \left(1 + \frac{p}{np_0} \right) Q_w - \frac{V}{np_0} \frac{dp}{dt} - \frac{Vp}{p_0} \frac{d(1/n)}{dt}, \tag{3.32}$$

• Exhalation

$$Q_p = Q_w - \frac{V}{n(p_0 + p)} \frac{dp}{dt} - \frac{Vp}{p_0 + p} \frac{d(1/n)}{dt}. \tag{3.33}$$

Equations (3.33) and (3.32) are essentially extended versions of the equations presented by [10], which take into account the effect on n of the real gas.

Efficiency is calculated as

$$\eta = \frac{P_w - P_{PTO}}{P_w}. \tag{3.34}$$

Figure 3.4 shows the efficiency calculated over a full wave cycle, including inhalation and exhalation stages, for a sinusoidal pressure signal of 100 kPa amplitude that approximates the harmonic behaviour of gas within an OWC device. The efficiencies during both exhalation and inhalation stages are lower for a real gas than a corresponding ideal gas. The discrepancy between the curves is less than 1.5% during

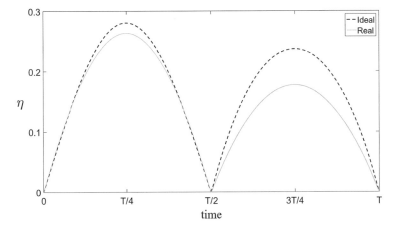

Fig. 3.4 Temporal variations of efficiency of an OWC covering the inhalation (first half of the cycle) and exhalation (second half of the cycle) stages over one oscillation period T, for a sinusoidal pressure signal of amplitude 100 kPa. (*Source* [6]. Reproduced with permission of Elsevier)

inhalation, and reaches about 6% during exhalation. This is caused by the presence of atmospheric air during inhalation. Conditions inside the chamber during this part of the cycle are less influenced by the conditions outside the chamber because the air is renewed and the air density can be considered as constant. During the first and third eighths of the cycle, the real gas efficiency curve slightly overlaps that of the ideal gas. This is again related to the presence of real air. While the chamber is filling with fresh air (at the beginning of inhalation), the real gas model takes into account the changes in density, and the process is then similar to that of an ideal gas. So long as the chamber is filled with fresh air (peak), the efficiency starts to vary as mixing occurs. During the second part of the inhalation process, the filling velocity starts to decrease, and the real and ideal gas power estimates converge according to Eq. (3.32).

A rough calculation is presented to demonstrate the effect of the real gas formulation on the predicted magnitude and phase of the airflow through the OWC turbine. Air flow in the OWC is calculated for exhalation and inhalation processes driven by harmonic pressure change using Eqs. (3.32) and (3.33) for real and ideal gas scenarios. The results are compared to predictions by [10] for an ideal gas flow. Figure 3.5 shows the time–dependent phase difference ($\Delta\psi$) obtained between the real and ideal air flow estimates over a complete pressure cycle. The curve exhibits almost no phase difference in the middle of the cycle because the pressure drop and associated water surface movement are both close to zero, and the ideal and real air flow models give the same results. As the pressure drop increases, the phase difference also increases, exhibiting a peak soon after the start of the cycle. The peak and trough represent the end of the expansion and compression process, and associated minimum and maximum water levels in the chamber. At $t = 0, T, 3T, etc.$ the water surface is alternately located at one of its extremes, thus explaining the large differ-

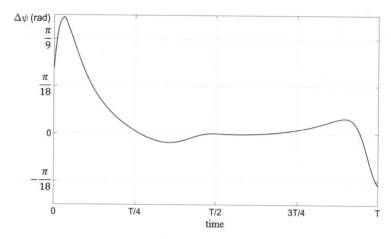

Fig. 3.5 Temporal variation in phase difference between real and ideal air flows for inhalation (first half of the cycle) and exhalation (second half) over a pressure oscillation period (T) in an OWC chamber. (*Source* [6]. Reproduced with permission of Elsevier)

ence between phases. Throughout the cycle, the phase gradually alters direction, with expansion and compression ending at opposite symmetric points. Note that if the two processes immediately succeed each other, then the first will start at its maximum, ending at zero, after which the second process starts at zero and ends at its minimum.

3.5   Conclusions

To date most analyses of the thermodynamic behaviour of an oscillating water column have been based on ideal gas theory. This chapter has extended the analysis to a real gas by deriving mathematical expressions for the adiabatic index which accounts for a water vapour-dry air mixture. The resulting index can be readily implemented in a numerical model, and is very straightforward to apply. Modified expressions have been developed for the specific heat coefficients, entropy, internal energy, enthalpy, chemical potential, and speed of sound, all of which apply to a real gas. Unlike ideal gas theory, the real gas theory is found to give excellent agreement with experimental data on mass flow conservation and energy balance through a laboratory-scale OWC. It is found that the adiabatic index depends linearly on temperature and pressure. The real gas theory has also been applied to the hypothetical OWC device proposed by [10], obtaining modified expressions for inhalation and exhalation of the air flow. It is shown that differences between the ideal and real gas models can reach 6%. The real gas model explains part of the losses observed in OWC plants, [11]. This study analyses the OWC chamber as an isolated system to understand the effects of the real gas formulation in a simple way. However, the changes introduced by the real gas model in the OWC chamber conditions affect the radiation coefficients as presented by [5]. This will modify the OWC–wave interaction patterns, and further variations are expected between ideal and real gas models. Moreover, it is observed that the accuracy of the real gas model to the actual OWC working conditions is mainly dependent on the adiabatic index n, particularly on changes in specific heat at constant pressure. A time-dependent specific heat is closer to a precise simulation of the real process. The present analysis should be particularly useful to engineering practitioners involved in the design of oscillating water columns, and may have more general application to turbines through which water vapour-dry air mixtures pass.

Acknowledgements Encarnación Medina has been recipient of a scholarship from the TALENTIA Fellowship Programme, funded by the Regional Ministry of Economy, Innovation, Science and Employment in Andalusia (Spain).

References

1. Biel Gayé, J.: Curso sobre el Formalismo y los Métodos de la Termodinámica. Reverté (1997). ISBN 9788429143430
2. Cengel, Y.A., Boles, M.A.: Thermodynamics. An engineering approach, McGraw-Hill Education (2015). ISBN 978-0-07-339817-4
3. El-Twaty, A.I., Prausnitz, J.M.: IGeneralized van der waals partition function for fluids. modification to yield better second virial coeffcients. Fluid Phase Equilib. **5**, 191–197 (1981)
4. Falcão, AFdO, Justino, P.A.P.: OWC wave energy devices with air flow control. Ocean Eng. **26**, 1275–1295 (1999)
5. Martins-Rivas, H., Mei, C.C.: Wave power extraction from an oscillating water column at the tip of a breakwater. J. Fluid Mech. **626**, 395–414 (2009)
6. Medina-López, E., Moñino, A., Borthwick, A., Clavero, M.: Thermodynamics of an OWC containing real gas. Energy **135**, 709–717 (2017)
7. Medina-López, E., Moñino, A., Clavero, M., Del Pino, C., Losada, M.A.: Note on a real gas model for OWC performance. Renew. Energy **85**, 588–597 (2016)
8. Plank, M.: Treatise on Thermodynamics. Dover Publications Inc. (1905). ISBN 978-0486663715
9. Prausnitz, J., Lichtenthaler, R., Gomes de Azevedo, E.: Molecular Thermodynamics of Fluid–Phase Equilibria. Prentice–Hall (1999). ISBN 0-13-977745-8
10. Sheng, W., Alcorn, R., Lewis, S.: On thermodynamics of primary energy conversion of OWC wave energy converters. Renew. Sustain. Energy **5**, 023 (105–1–17) (2013)
11. Trust, T.C.: Oscillating water column wave energy converter evaluation report. Tech. rep, Marine Energy Chall. (2005)
12. Tsonopoulos, C., Heidman, J.L.: From the virial to the cubic equation of state. Fluid Phase Equilib. **57**, 261–276 (1990)
13. Wisniak, J.: Eike kamerlingh-the virial equation of state. Indian J. Chem. Technol. **10**, 564–572 (2003)

Chapter 4
Numerical Simulation of an Oscillating Water Column Problem for Turbine Performance

Abstract Air turbines are commonly used in Oscillating Water Column (OWC) devices for wave energy conversion. This chapter presents a proposed methodology to simulate the performance of an OWC turbine through the implementation of an Actuator Disk Model (ADM) in Fluent®. A set of different regular wave tests are developed in a 2D numerical wave flume. The model is tested using the information analysed from experimental tests on a Wells type turbine, carried out in wind tunnel. Linear response is achieved in terms of pressure drop and air flow in all cases, proving effectively the actuator disk model applicability to OWC devices.

4.1 Objective

The aim of this chapter is to study the performance of an offshore OWC converter in a 2D numerical flume, accounting for the implementation of the linear response of the PTO system, i.e. a Wells turbine. The proposed model represents a simplified case in which the main features of the wave–device interaction can be visualized without loss of generality. Regarding the turbine performance simulation, the numerical model is adjusted following the information provided by the experimental study of a Wells turbine case study. The proposed methodology can be considered as a basis for the study of the OWC performance under average wave conditions in relatively calm coasts. Those conditions could might be predominant at potential deployment locations for off–shore wave energy converters, designed to work in intermediate to shallow—water areas —i. e. the Mediterranean and Atlantic coastlines in Southern Spain. The chapter is organized as follows: first, the methodology to set ready the numerical model is explained. Second, the results for hydrodynamic variables are presented, verifying the linear response of the model along with a comparison with experimental data. Finally, the discussion and conclusions are presented. The validation of the proposed numerical model for wave generation in FLUENT® is presented in Appendix I. The basic theory underlying the numerical modelling procedure is explained in Appendix II.

© The Author(s) 2018
A. Moñino et al., *Thermodynamics and Morphodynamics in Wave Energy*,
SpringerBriefs in Energy, https://doi.org/10.1007/978-3-319-90701-7_4

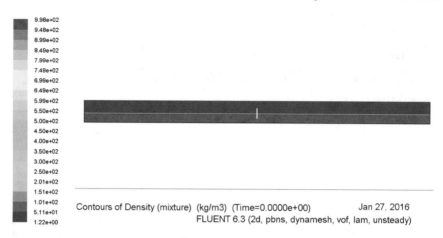

Fig. 4.1 Numerical flume in FLUENT®. Paddle to the left of OWC device. (*Source* [19]. Reproduced with permission of Elsevier)

4.2 Methodology

4.2.1 General

A simple off–shore OWC converter is implemented in a numerical flume in FLUENT®. The problem is simplified to represent an infinite wide OWC slot subsequently represented by a 2D domain. No further deepening in the geometrical factors of the OWC and chamber configuration has been considered in this work.

The 2D flume is 200 m long and 12 m high with 5 m depth, see Fig. 4.1. The OWC is placed at the centre of the domain and submerged 2.5 m. The detailed scheme of the OWC is depicted in Fig. 4.2. The flume is designed to be threefold the wavelength. Although that value is slightly smaller than other proposed models, [21], it suffices to visualize the basic phenomena concerned in this chapter without loss of generality.

Regarding the power take–off simulation, different numerical models can be used to represent the dynamics of flow through a turbine, e.g. Single Rotating Reference Frame (SRF), Virtual Blade Model (VBM) and Actuator Disk Model (ADM), [7, 15]. Those methods have proven to be totally suitable for their purposes. The first two models are very accurate, and present very good results related to velocity field measurements, [15]. The last one is not of straightforward application when studying the velocity field around blades. However, if the main goal is to identify control volume relations between inlet velocity and pressure drop instead of turbine dynamics (e.g. drag and lift forces on blades, axial and radial velocity components through vanes, air tangential velocity between blade tip and external hub, etc.), then the ADM is an accurate option, [11]. Therefore, the proportionality between flow discharge and pressure drop in the turbine response is achieved through the implementation of an ADM configuration. The authors in this chapter have used the ADM because it

Fig. 4.2 OWC model.
(*Source* [19]. Reproduced
with permission of Elsevier)

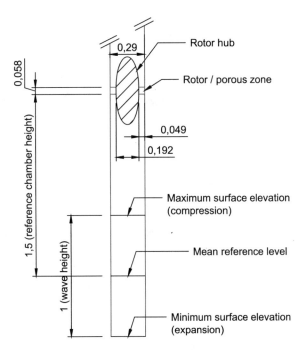

provides a realistic physical representation of the turbine. This is represented as a region with the same thickness and pressure drop as the real turbine, in which effects of inertial and viscous resistance can be accurately controlled. In fact, the ADM theory was formerly intended to represent the redistribution of mass flow outside the rows of blades in axial flow machines, [20]. In addition, it must be kept in mind that the purpose of this chapter is to provide a built–in methodology to accurately simulate the OWC compression/expansion cycle in any simulation domain, rather than to compare the quality of specific features representing some mechanisms involved in parts of the complete cycle.

In the ADM, the aerodynamic effect of rotating blades is described by a pressure discontinuity over a thin disk with a cross sectional area equal to the swept area of the rotor. Therefore the thin disk is modelled as a porous media that induces a pressure drop in the flow, but with continuous distribution of the velocity field, see Appendix II.

4.2.2 Model Set up and Wave Generation

The numerical domain is meshed in GAMBIT® under a *Tri/Pave* scheme with maximum spacing setting of 0.1 m and a minimum element size of 0.001 m, see Fig. 4.3. To help obtaining a smoother structure in the final mesh, a pre–meshing scheme is applied to the hub edges inside the chamber, with spacing of 0.01 m. As a result,

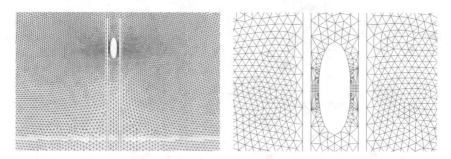

Fig. 4.3 Detailed view of OWC device in the numerical flume (left) and mesh detail around porous layer (right). (*Source* [19]. Reproduced with permission of Elsevier)

a mesh with 221578 elements is generated. The FLUENT® solver is configured to *laminar* and *VOF* (Volume of Fluid) with air as phase 1 and water as phase 2. The pressure–velocity coupling is set to *PISO* (Pressure Implicit Split Operator) and the discretization scheme for pressure is set to *PRESTO* (Pressure Staggering Options), while for momentum equations is set to 1st *order upwind*. The free surface reconstruction is set to *Geo-Reconstruct*.

Different approaches can be used to generate the wave field in the simulation domain. Methodologies have been successfully developed by some authors, [8, 12, 14], obtaining accurate results for regular waves. In this chapter, the dynamic mesh scheme implemented in FLUENT® has been used for the purpose. It provides an efficient wave generation in terms of solver time and deforming cells. The calibration of the wave generation shows good results when compared with wave flume data, as it is shown in Appendix I. The moving paddle scheme generates a wave field as close as possible to the one observed in an experimental flume.

Waves are generated in the flume by a piston–type paddle located at the left side of the flume. The paddle motion is enabled by a dynamic mesh scheme assigned to the paddle boundary. For the purpose of the study, regular waves are generated up to 2 m height and 8 s period according to low–average climate conditions, with intermediate water depth propagation conditions following previous reference cases, [21]. No irregular waves are considered.

The paddle motion is accomplished by a dynamic mesh scheme. This model is validated against experimental data as shown in Appendix I. Time–dependent variables determining the instantaneous paddle position and velocity are implemented in a compiled UDF in the model, following the Bisel theory for a piston–type paddle, [13]:

$$x_{paddle} = \frac{S_0}{2} \cos \frac{2\pi t}{T}$$

$$U_{paddle} = -\frac{S_0 \pi}{T} \sin \frac{2\pi t}{T} \tag{4.1}$$

$$S_0 = H \frac{2k_0 h + \sinh 2k_0 h}{4 \sinh^2 k_0 h}$$

where x_{paddle} is the paddle displacement along the horizontal axis, U_{paddle} is the paddle displacement velocity, S_0 is the paddle stroke, H is the wave height, k is the wave number, h is the water depth at the paddle location, T is the wave period and t is the simulation time. The dynamic mesh ensures the harmonic oscillation of the free surface described by:

$$\begin{aligned}
h_{owc} &= h_{ref} - \eta \\
V_{owc} &= S_{owc}(h_{ref} - h_{owc}) \\
U_{owc} &= \frac{d\eta}{dt} = -\frac{H\pi}{T}\sin\frac{2\pi t}{T}
\end{aligned}$$
(4.2)

4.2.3 Porous Zone Configuration for Turbine Performance Simulation

The turbine performance is represented by means of the actuator disk model. The porous medium involved in the ADM definition is modeled by the addition of a momentum source term to the standard fluid flow equations. The source term consists of two parts: a viscous loss term (the first term on the right-hand side of equation (4.3)), and an inertial loss term (the second term on the right-hand side of Eq. (4.3)), [9]:

$$\delta S_i = -\delta \left(\sum_{j=1}^{3} D_{ij} \mu v_j + \sum_{j=1}^{3} C_{ij} \frac{1}{2} \rho |v| v_j \right)$$
(4.3)

where S_i is the source term for the momentum equation, δ is the porous layer thickness, $|v|$ is the magnitude of the velocity and D and C are prescribed matrices. This momentum source contributes to the pressure gradient in the porous cell, creating a pressure drop that is proportional to the fluid velocity (or velocity squared) in the cell. The units for the momentum source are N/m^3, so this term has to be multiplied by the porous layer thickness to insert it as a component of the external body forces in the momentum conservation equations, [9].

Equation (4.3) can be adapted for a simple homogeneous porous media:

$$S_i = -\left(\frac{\mu}{\alpha} v_i + C_2 \frac{1}{2} \rho |v| v_i \right)$$
(4.4)

where α is the permeability and C_2 is the inertial resistance factor.

The porous zone settings are the same for air and water phases, accounting for the fact that water never reaches the porous zone for the generated waves. The inertial resistance coefficient is set to $C_2 = 0$ for both x —horizontal— and y — vertical— directions. The viscous resistance coefficient is set to $1/\alpha = 10^7 \, m^{-2}$ in the y direction, while for the x direction is set to $1/\alpha = 10^{10} \, m^{-2}$ in order to prevent undesired radial flow effects. The viscous resistance coefficients are adjusted following the measurements for a Wells turbine in a wind tunnel, see Figs. 4.4 and 4.5, [2]. The performance of the experimental turbine model shows a linear dependence between pressure drop and flow rate. For flow rates under $0.26\,\mathrm{m^3/s}$ a stall effect seems to be present and the turbine is not self–starting, as will be discussed in Sect. 4.3.3.

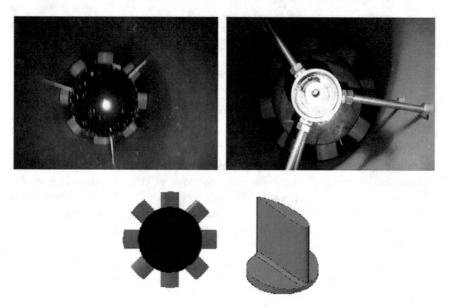

Fig. 4.4 Photography of the experimental turbine inlet (top left) and outlet (top right). 3D composition of turbine and blade (bottom). (*Sources* [2] and [19]. Reproduced with permission of Elsevier)

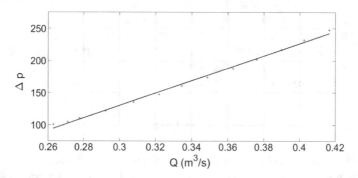

Fig. 4.5 Pressure drop versus flow velocity for the eperimetal turbine. (*Sources* [2] and [19]. Reproduced with permission of Elsevier)

The adjustment of coefficients ensures the same relationship between pressure drop and flow velocity as in the experiments. The geometry for the design of the turbine in terms of external diameter, hub design and hub to tip radio is the same as in the numerical model. The experimental turbine characteristics are summarized in Table 4.1 and sketched in Fig. 4.6. The turbine has 8 right blades composed by profiles of an airfoil *NACA0018* with constant cord length.

Table 4.1 Experimental turbine characteristics

Blade airfoil	NACA0018
D_t	0.3 m
c	52.2 mm (constant)
N_b	8
R_h	96 mm
R_t	144 mm
$h = R_h/R_t$	0.67
σ	0.55

(*Source* [2])

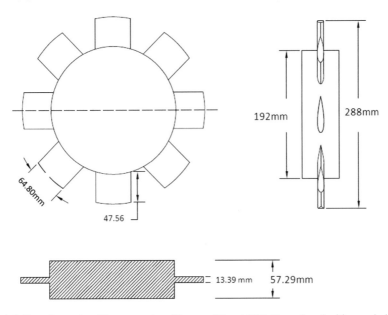

Fig. 4.6 Experimental turbine geometry. (*Sources* [2] and [19]. Reproduced with permission of Elsevier)

4.3 Results and Discussion

4.3.1 Surface Elevation, Wave Spectra, Pressure Drop and Air Velocity

Regular waves are run for tests: (a) $H = 0.5$ m, $T = 3.5$ s; (b) $H = 1$ m, $T = 6$ s; (c) $H = 1.5$ m, $T = 7$ s and (d) $H = 2$ m, $T = 8$ s. The gauge distribution inside the flume is depicted in Fig. 4.7. Surface elevation time series in the flume and inside the OWC are represented in Fig. 4.8 for the different tests. Although tests were run to record at least 5 waves in each case, results are plotted for the first 25 s in order to compare different results. Data are recorded every 0.1 s. Surface elevation in every sensor for the complete simulation time is shown in Fig. 4.9 for the case $H = 1$ m, $T = 6$ s, in order to provide with an overview of the differences between sensor measurements. The wave generation addresses a regular pattern and no second order effects are observed.

The water surface elevation time series measured at gauges $S1$, $S2$ and $S3$ are used for the calculation of the complex reflection coefficient modulus and phase, and for the incident wave separation. The incident and reflected wave trains are separated by applying the method described in [1], according to Linear Wave Theory. This method is based on the three–gauge method in [17], but it resolves the mathematical inconsistence of minimizing the complex variable with which it was formulated. Following the calculation and accounting for the specific flume set–up, the modulus of the reflection coefficient (K_R) is calculated, see Table 4.2. Values of K_R range from ~0.2 up to 0.4 for the highest period. In addition, the ratio between the external OWC diameter and the wavelength is $D/L < 0.02 << 0.2$, revealing that back scattered waves could have a minor effect for higher kh, [3], explaining the lower reflection values in those.

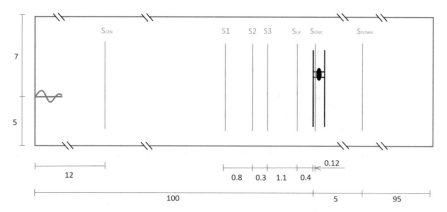

Fig. 4.7 Numerical set–up and gauge scheme. (*Source* [19]. Reproduced with permission of Elsevier)

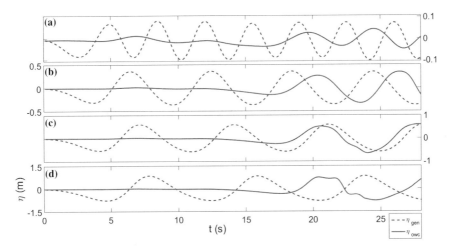

Fig. 4.8 Surface elevation (leeward the paddle and inside OWC) for tests: **a** $H = 0.5\,\text{m}$, $T = 3.5\,\text{s}$; **b** $H = 1\,\text{m}$, $T = 6\,\text{s}$; **c** $H = 1.5\,\text{m}$, $T = 7\,\text{s}$ and **d** $H = 2\,\text{m}$, $T = 8\,\text{s}$. (*Source* [19]. Reproduced with permission of Elsevier)

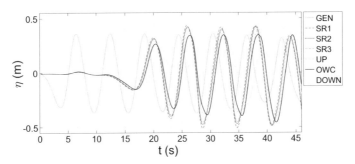

Fig. 4.9 Surface elevation for test $H = 1\,\text{m}$, $T = 6\,\text{s}$ in all sensors. (*Source* [19]. Reproduced with permission of Elsevier)

The proposed model allows to perform some basic calculations regarding the energy budget, which is a first approach for the comprehensive interpretation of the theoretical solutions for the radiation–diffraction problem. The spectral density can be calculated for a given wave record as:

$$S_f = \frac{A^2/2}{\delta f} \tag{4.5}$$

where A is the wave amplitude for each frequency interval δf. For the incident waves on the seaward side of the OWC and inside the converter, the spectral densities are represented in Figs. 4.10 and 4.11. The wave energy can be estimated from the area under the spectra:

$$E = \rho g \int S(f)df \qquad (4.6)$$

The energy impinging the OWC and the available energy inside the device due to the water column oscillation can be calculated in that way. The results are summarized in Table 4.2 in terms of kh, where k is the wave number and h is the water depth. E represents the energy of the impinging waves outside the OWC. Available wave energy inside the OWC (E_{OWC}) equals the energy on the seaward side of the device

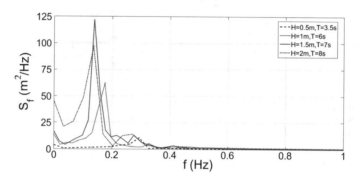

Fig. 4.10 Wave spectra for tests: **a** $H = 0.5$ m, $T = 3.5$ s; **b** $H = 1$ m, $T = 6$ s; **c** $H = 1.5$ m, $T = 7$ s and **d** $H = 2$ m, $T = 8$ s. (*Source* [19]. Reproduced with permission of Elsevier)

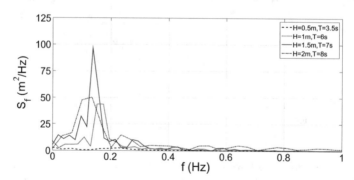

Fig. 4.11 Wave spectra inside the OWC for tests: **a** $H = 0.5$ m, $T = 3.5$ s; **b** $H = 1$ m, $T = 6$ s; **c** $H = 1.5$ m, $T = 7$ s and **d** $H = 2$ m, $T = 8$ s. (*Source* [19]. Reproduced with permission of Elsevier)

Table 4.2 Wave energy impinging the OWC and inside the converter

Test	kh	K_R	E (MJ/m²)	E_{OWC} (MJ/m²)	E_{OWC}/E (%)
$H = 0.5$ m; $T = 3.5$ s	1.74	0.30	0.41	0.2	48.2
$H = 1$ m; $T = 6$ s	0.83	0.20	2.44	1.93	79.2
$H = 1.5$ m; $T = 7$ s	0.69	0.16	3.74	3.57	95.5
$H = 2$ m; $T = 8$ s	0.59	0.40	3.21	3.21	100.0

for the two highest periods, rendering a ratio of almost 100 %. That is in accordance with the higher water elevations reached inside the chamber for those cases, see Fig. 4.8. It must be recalled that this ratio only represents the availability of the resource inside the OWC in reference to the resource outside for incoming waves during a time interval, i.e. the amount of energy to be seized by the OWC for that complete interval, which for the highest period could be expected to be essentially of 100 %. Moreover, this value is only possible because this is a 2D simulation. In a 3D case, without an infinite-width OWC a 100 % energy capture would never come up. Obviously, the expected OWC performance would be lower than the calculated ratio, due to the delay between the external oscillation and the water surface displacement inside the chamber, as it can be observed in Fig. 4.8. In fact, the theoretical solutions of the radiation–diffraction problem for the OWC provide values of surface elevation, amplitude and phase for a given configuration. Therefore, the numerical simulations provide a first visualization of the wave interaction with the device in an appropriate way.

The pressure drop and velocity inside the OWC are plotted in Fig. 4.12. Pressure drop and vertical velocity values inside the OWC chamber exhibit the same pattern. Pressure drop increases with wave height and period, in correspondence to the increase in the velocity of the surface oscillation.

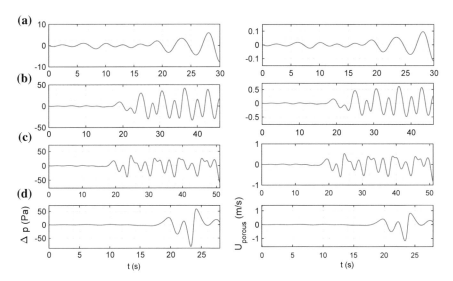

Fig. 4.12 Pressure drop inside the OWC chamber (left) and flow velocity at the inlet of the porous region (right) for tests: **a** $H = 0.5\,\text{m}$, $T = 3.5\,\text{s}$; **b** $H = 1\,\text{m}$, $T = 6\,\text{s}$; **c** $H = 1.5\,\text{m}$, $T = 7\,\text{s}$ and **d** $H = 2\,\text{m}$, $T = 8\,\text{s}$. (*Source* [19]. Reproduced with permission of Elsevier)

4.3.2   Pressure Drop Versus Air Flow Through the Converter

Results on the generated wave height and the pressure drop versus flow plots are shown in Fig. 4.13.

The pressure drop is linear with the air flow through the porous zone, as shown in Fig. 4.13. The pressure range is consistent with the expected values: higher values of pressure and flow rate corresponds with higher wavelength and period for the incoming waves. It is important to note that the linearity is conserved even with the

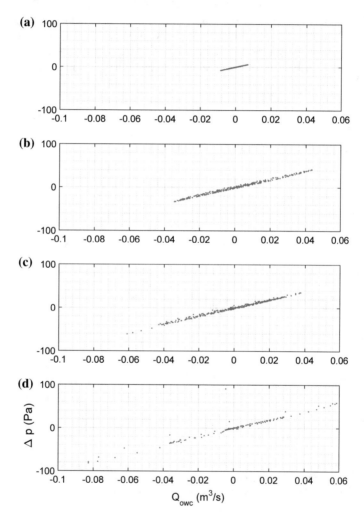

Fig. 4.13 Pressure drop versus flow velocity for tests: **a** $H = 0.5\,\mathrm{m}, T = 3.5\,\mathrm{s}$; **b** $H = 1\,\mathrm{m}, T = 6\,\mathrm{s}$; **c** $H = 1.5\,\mathrm{m}, T = 7\,\mathrm{s}$ and **d** $H = 2\,\mathrm{m}, T = 8\,\mathrm{s}$. (*Source* [19]. Reproduced with permission of Elsevier)

Fig. 4.14 Relative effects of
hub-to-tip ratio and solidity
on the self–start condition of
the turbine, [20]. (*Source*
[19]. Reproduced with
permission of Elsevier)

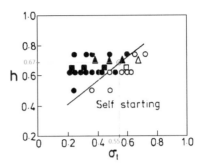

bigger waves. The central point of the Δp-Q_{owc} graph is zero, as there is not any kind of initial resistance imposed to the ADM. Moreover, friction is not a factor included in this tests, which could deviate the central point of the plots showed in all previous results (Fig. 4.14).

4.3.3 Turbine Performance Simulation by the ADM

The comparison between the pressure drop-air flow relationship obtained with the experimental Wells turbine and the ADM is shown in Fig. 4.15. In order to compare experiments and ADM results, if the initial resistance effects were neglected, the translation of the initial experimental line is done taking into account the minimum velocity necessary to allow the Wells turbine rotation. This velocity is set to zero ($v(p = 0) = 0$) to simulate the ADM zero-resistance performance, and the line is translated in consequence. The result is the superposition of the experimental and numerical results, so the ADM operation in this case would be validated, if dynamic effects as pointed above are neglected. In any case, the numerical model reproduces the linear response of the turbine in terms of pressure and air flow according to the experimental data.

The reasons for the difference observed between the experimental turbine and the ADM model might be related with several effects in which mechanical friction, dynamic drag and start up characteristics, or *crawling* phenomenon, are included. Form a general point of view, the air passing through the turbine has to overcome a mechanical and friction resistance, so an increment in air flow—or if preferred, in pressure—is going to appear, since a fraction of the flow energy is used to overcome that permanent resistance. This fact could lead the parallelism between the Δp-Q_{owc} plots in both experimental and numerical models. As the ADM is not designed to reproduce dynamic effects, this reaction cannot be taken into account in the numerical model, i.e. the ADM resistance to air flow is null. In fact this is why the ADM Δp-Q_{owc} plots are always centred in ($Q_{owc} = 0$, $\Delta p = 0$), a condition that might not be applicable to real Wells turbines.

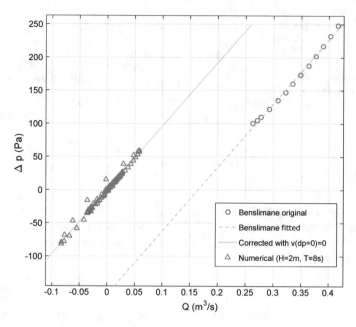

Fig. 4.15 Comparison between experiments in physical tunnel and numerical results. Pressure drop versus flow velocity. Note that the corrected line is the Benslimane fitted line translated to remove the offset. (*Source* [19]. Reproduced with permission of Elsevier)

In addition, the *crawling* phenomenon is generally observed in Wells turbines, and occurs when the turbine is not able to achieve the operation velocity by itself from the idle state. Turbine *crawling* depends on two main factors:

- the turbine *solidity* σ.
- the relationship between the *hub-to-tip ratio* ($h = R_h/R_t$) and the solidity.

Following Fig. 4.14, and taking into account the turbine hub-to-tip ratio is $h = 0.67$ and the turbine solidity is $\sigma = 0.55$, the experimental turbine is on the threshold of not being self starting. Obviously, that effect is not represented in the porous zone performance.

4.4   Conclusions

This chapter presents the results of the study on the performance of an off–shore OWC converter in a 2D flume, in which it is implemented the linear response of the PTO system, usually consisting of a Wells turbine. The chapter follows the research line started in [18], which attempts to advance in the knowledge on the influence of moist air in OWC energy extraction efficiency. A series of regular waves tests are carried out and compared with the experiments developed by [2] to calibrate the model.

The linear response of the turbine has been achieved through the application of the ADM theory. The proposed methodology provides with coherent results regarding pressure drop and air flow rates through the OWC. The main conclusions of this research are:

1. The *actuator disk model* theory applies effectively for a turbine with linear response in terms of pressure drop and air flow rate under dry air conditions, i.e. a Wells turbine. A series of regular waves tests were run in a 2D numerical flume for the purpose, showing a good response of the ADM independently of the wave heights and periods tested.
2. A simplified case study is proposed. The numerical set–up provides consistent results regarding the basic principles in the radiation–diffraction problem and the wave–OWC interaction. The linear response of the PTO as represented by the ADM is achieved with all test cases.
3. Accounting for the limitation of the ADM in terms of the feasibility to represent the turbine dynamics, the porous zone parameters can be adjusted to ensure the correct performance in terms of pressure drop and flow rate.
4. Although FLUENT® software has been used for this work, the methodology can be implemented through other finite element solver software.

4.5 Future Research

The actuator disk model applied to OWC has demonstrated to be a good tool to simulate the Wells turbine performance. Now that this tool has been tested for regular conditions with dry environment, a further step could be done following [18] and implement the effect of moisture in the OWC chamber.

Moreover, this chapter sets the basic conditions for a further research in *System Identification* tools applied to wave energy extraction. Those tools represent an affective method to study the OWC performance in terms of radiation-scattering features under certain considerations. The research lines presented by [5, 10, 16] are excellent as a start point to apply *System Identification* tools to OWC.

Future research should focus on the analysis of compressibility effects for prototype scale, following the line of [6]. Moreover, an active absorption capability should be inserted in the model and compared with the actual one to set the differences and effects of this tool in the simulation of linear turbines.

Appendix I: Numerical Model Validation

The numerical model for wave generation in FLUENT® used in this chapter has been validated against experimental data. The experiments were carried out in a wave flume at the *Andalusian Institue for Earth System Research, Universidad de Granada* (Spain).

Figure 4.16 shows the experimental set up. A 23 m long wave flume is presented, with a dissipation beach at the right end, and a piston–type paddle located on the left. The water depth is 0.4 m and the total height of the flume is 0.8 m. Three wave gauges are located in the central section of the flume to analyse incident and reflected waves. For the purpose, a 2D numerical flume, see Figs. 4.17 and 4.18, has been meshed in GAMBIT® and configured in FLUENT®, to check the validity of the numerical model. The mesh has a total of 43544 quadrilateral 0.01 m length sided cells. Finer meshes have been checked in order to verify the accuracy in the results, but no significant differences have been observed for the purpose in terms of wave propagation and surface elevation detection. Hence, for the sake of run time efficiency a coarser mesh has been chose. In any case, the solver configuration has been set following the same scheme as described in Sect. 4.2.2.

Table 4.3 summarizes the set of wave heights and wave periods configured for the numerical validation, following experimental test cases.

The results show good agreement between experiments and numerical model, see Figs. 4.19 and 4.20. At the beginning of the simulation there is a slight out of phase effect between numerical and experimental waves, see Fig. 4.19. The reason for that mismatch lies on the experimental paddle gain run up, which is not present in the

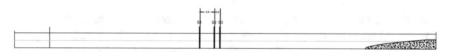

Fig. 4.16 Experimental set up for numerical model validation. (*Source* [19]. Reproduced with permission of Elsevier)

Fig. 4.17 Finite element domain in the numerical flume. (*Source* [19]. Reproduced with permission of Elsevier)

Fig. 4.18 Detailed mesh in the dissipative beach region. (*Source* [19]. Reproduced with permission of Elsevier)

Table 4.3 Tests run for model validation

H (m)	T (s)
0.113	1.5
0.052	3

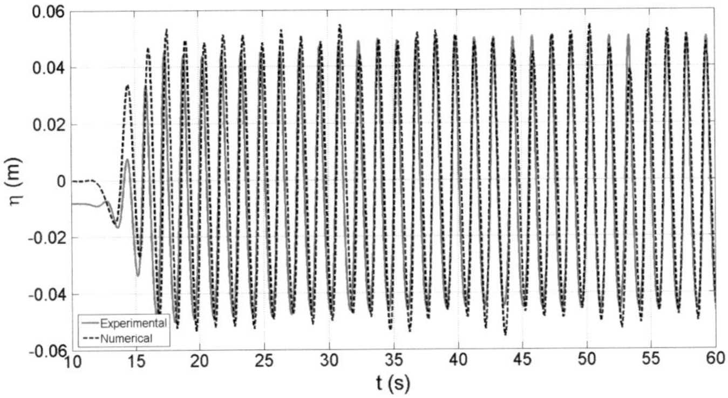

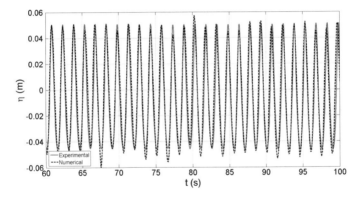

Fig. 4.19 Experimental versus numerical wave height. $H = 0.113$ m, $T = 1.5$ s. First part of the test. (*Source* [19]. Reproduced with permission of Elsevier)

Fig. 4.20 Experimental versus numerical wave height. $H = 0.113$ m, $T = 1.5$ s. Last part of the test. (*Source* [19]. Reproduced with permission of Elsevier)

numerical paddle. In any case, numerical and experimental wave heights match in phase and height with time advancement, see Fig. 4.20.

Same comments for test $H = 0.052$ m, $T = 3$ s, see Fig. 4.21. However, in the last part of the test, Fig. 4.22, some peaks are present every three waves. The magnitude of these peaks is around 0.005 m, which can be simply due to the surface elevation calculations. This phenomenon is present for very little wave heights, in tests run over 0.1 m this is not observed. The authors think that this fact is present because of the size of the water–air inter-phase in the numerical tests, which is close to the

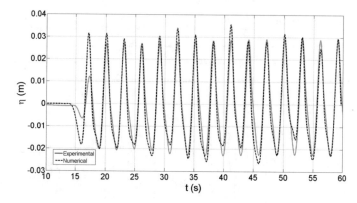

Fig. 4.21 Experimental versus numerical wave height. $H = 0.052\,\text{m}$, $T = 3\,\text{s}$. First part of the test. (*Source* [19]. Reproduced with permission of Elsevier)

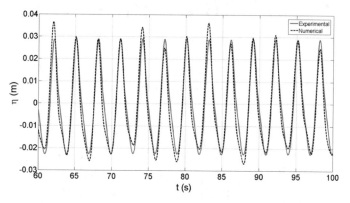

Fig. 4.22 Experimental versus numerical wave height. $H = 0.052\,\text{m}$, $T = 3\,\text{s}$. Last part of the test. (*Source* [19]. Reproduced with permission of Elsevier)

wave height in this case. That would explain the mismatch between numerical and experimental results in this test.

Please note that the authors have not included any kind of active absorbing capability. However, results show a good agreement with experiments, and thus the result are considered to be valid for the purpose of this chapter, which focuses on the simulation of a turbine. However, this fact should be considered in future analyses.

Appendix II: Basic Theory of the ADM

The actuator disk is conceived as an infinitesimal thin disk located perpendicular to the flow. The ADM produces a change in the pressure field across while preserving a soft variation in the velocity field. In other words, the disk extracts the pressure

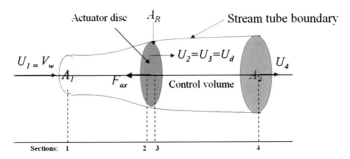

Fig. 4.23 3D actuator disk model schematic. (*Sources* [22] and [19]. Reproduced with permission of Elsevier)

energy but does not affect the kinetic energy, [4]. As the fluid approaches the disk, the pressure increases upstream while the fluid velocity slows down, see Fig. 4.23.

Inside the stream tube surrounding the disk the mass, momentum and energy conservation are applied. The flow assumptions are: homogeneous, incompressible and inviscid fluid, steady state flow and uniform flow conditions over the disk. So the three equations of conservation are:

$$\rho U A = \rho_1 U_1 A_1 = \rho U_d A_R = \rho U_4 A_4 = \dot{m} \qquad (4.7)$$

$$\dot{m}(U_1 - U_4) = T + F_x \qquad (4.8)$$

$$\dot{m} = \left(\frac{U_1^2}{2} - \frac{U_4^2}{2}\right) = T U_d \qquad (4.9)$$

So, the induction velocity U_d can be obtained from Eqs. (4.8) and (4.9) as the semi-sum of the inlet and outlet velocities, U_1 and U_4, respectively. The axial induction factor is:

$$a = \frac{u_i}{V_w} \qquad (4.10)$$

where u_i is the axial component of the induction velocity U_d.

Hence, thrust and power can be defined as:

$$T = \rho V_w^2 2a(1-a) A_R \qquad (4.11)$$

$$P = \rho V_w^3 2a(1-a)^2 A_R \qquad (4.12)$$

and the thrust and power coefficients are as follows. A comparison between them can be seen in Fig. 4.24.

Fig. 4.24 Graph of a in front of C_T and C_P. (*Sources* [4] and [19]. Reproduced with permission of Elsevier)

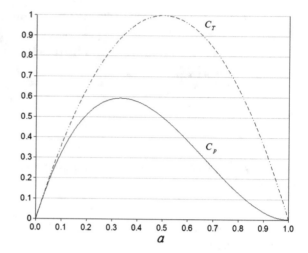

$$C_T = \frac{T}{1/2\rho V_w^2 A_R} = 4a(1-a) \tag{4.13}$$

$$C_P = \frac{P}{1/2\rho V_w^3 A_R} = 4a(1-a)^2 \tag{4.14}$$

References

1. Baquerizo, A.: Reflexin del oleaje en playas: Mtodos de evaluacin y prediccin. Ph.d. dissertation, Universidad de Cantabria, Santander, Spain (1995)
2. Benslimane, S.: Experimental study of the effect of blade sweep on the performance of wells turbine. University of Granada, Tech. rep. (2013)
3. Paixão Conde, J.M., Teixeira, P., Didier, E.: Numerical simulation of an oscillating water column device using a code based on Navier–stokes equations. In: Proceedings of the XXI International Offshore and Polar Engineering Conference (Maui, Hawaii), pp. 668–674 (2011)
4. Creech, A.: A three-dimensional numerical model of a horizontal axis, energy extracting turbine. An implementation on a parallel computing system. Ph.D. thesis, Heriot–Watt University, Scotland (2009)
5. Davidson, J., Giorgi, S., Ringwood, V.: Linear parametric hydrodynamic models for ocean wave energy converters identified from numerical wave tank experiments. Ocean Eng. **103**, 31–39 (2015)
6. Dimakopoulos, A., Cooker, M., Bruce, T.: The influence of scale on the air flow and pressure in the modelling of oscillating water column wave energy converters. Int. J. Mar. Energy **19**, 272–291 (2017)
7. Dimakopoulos, A., Cooker, M., Medina-Lopez, E., Longo, D.: Flow characterisation and numerical modelling of OWC wave energy converters. In: European Wave and Tidal Energy Conference (EWTEC, Nantes (France)) (2015)

8. Dimakopoulos, A., Cuomo, G., Chandler, I.: Optimized generation and absorption for three-dimensional numerical wave and current facilities. J. Waterway Port Coastal Ocean Eng. **06016001** (2016)

9. Fluent, I.: FLUENT 6.3 Users guide −7.19.6 user inputs for porous media. Fluent Inc., Centerra Resource Park, 10 Cavendish Court, Lebanon, NH 03766 (2006)

10. Gkikas, G.D., Athanassoulis, D.A.: Development of a novel non-linear system identification scheme for the pressure fluctuation inside an oscillating water column wave energy converter Part I: theoretical background and harmonic excitation case. Ocean Eng. **80**, 84–89 (2014)

11. Hamill, G., Kee, C.: Predicting axial velocity profiles within a diffusing marine propeller jet. Ocean Eng. **124**(15), 104–112 (2016)

12. Higuera, P., Losada, I.J., Lara, J.L.: Three-dimensional numerical wave generation with moving boundaries. Coastal Eng. **101**, 35–47 (2015)

13. Hughes, S.A.: Physical models and laboratory techniques in coastal engineering, vol. 7. World Scientific (1993)

14. Jacobsen, N.G., Fuhrman, D.R., Fredsoe, J.: A wave generation toolbox for the open-source cfd library: Openfoam. Int. Numer. Methods Fluids **70**(9), 1073–1088 (2012)

15. Javaherchi Mozafari, A.T.: Numerical modeling of tidal turbines: methodology development and potential physical environmental effects. Ph.D. thesis, University of Washington (2010)

16. López, I., Iglesias, G.: Efficiency of owc wave energy converters: a virtual laboratory. Appl. Ocean Res. **44**, 63–70 (2014)

17. Mansard, E.P.D., Funke, E.R.: On the Reflection Analysis of Irregular Waves. Laboratory Technical Report TR-HY-O17. Tech. rep., Natl. Res. Counc. Rev. Can. Hydraulics (1987)

18. Medina-López, E., Moñino, A., Clavero, M., Del Pino, C., Losada, M.A.: Note on a real gas model for OWC performance. Renew. Energy **85**, 588–597 (2016)

19. Moñino, A., Medina-López, E., Clavero, M., Benslimane, S.: Numerical simulation of a simple OWC problem for turbine performance. Int. J. Mar. Energy **20**, 17–32 (2017)

20. Raghunathan, S.: The wells turbine for wave energy conversion. Prog. Aerosp. Sci. **31**, 335–386 (1995)

21. Teixeira, P., Davyt, D., Didier, E., Ramalhais, R.: Numerical simulation of an oscillating water column device using a code based on navier-stokes equations. Energy **61**, 513–530 (2013)

22. Van Bussel, G.: Wind Energy Online Reader. TU Delft University (2008)

Chapter 5
Effects of Seabed Morphology on Oscillating Water Column Wave Energy Converter Performance

Abstract This chapter presents a numerical model to analyse the effects of changes in the bedforms morphology on Oscillating Water Column (OWC) wave energy devices. The model was developed in FLUENT® and based on the Actuator Disk Model theory to simulate the turbine performance. The seabed forms were reproduced with the morphodynamic model XBeach-G for a series of characteristic sea states in Playa Granada (southern Spain). These bedforms were used as input bed geometries in FLUENT® and compared with a hypothetical flat seabed to analyse the effects of changes in bed level on the OWC performance. Results of the simulated sea states reveal the influence of the seabed morphology in the power take–off performance, affecting the relationship between pressure drop and air flow rate through the turbine. Energy dissipation was found to be directly dependent on the bedforms unit volume. This lead to lower mean efficiencies for the cases with evolved morphologies (up to 15%) compared to those obtained for the hypothetical flat cases (19%). The effects of seabed formations on the power take–off performance presented in this chapter can be of interest in planning control strategies for OWC devices.

5.1 Objective

The main objective of this chapter is to study the interaction of a single OWC device with the coastal morphology, and to understand the effects of the seabed forms on the general OWC performance. The study cases are based on wave conditions in a gravel-dominated beach (Playa Granada, southern Spain), whose dynamics has been widely studied, characterized and modelled in recent years [1–3, 5–9]. It is a Mediterranean deltaic coast bounded to the east by Motril Port, which contributes to the generation of bedforms in front of the reflective breakwater [16]. Therefore, this site represents a valuable prototype to address the implications of bedforms on the OWC performance.

© The Author(s) 2018
A. Moñino et al., *Thermodynamics and Morphodynamics in Wave Energy*,
SpringerBriefs in Energy, https://doi.org/10.1007/978-3-319-90701-7_5

5.2   Methodology

The methodology followed in this chapter concerns both the selection of the characteristic wave cases, and the numerical set up for the simulation of the seabed evolution and OWC performance under wave action. The case presented is ideal, but based on a real problem. On the one hand, the numerical mesh, OWC design and case structure is taken from [15], where an Actuator Disk Model that simulates a Wells turbine is presented. On the other hand, sea states are based on wave conditions presented in Playa Granada (southern Spain).

5.2.1   Wave Case Selection

The wave conditions cases were selected based on a 58–year series of hourly data, corresponding to SIMAR point number 2041080 and provided by *Puertos del Estado*. Eleven values (one every 0.5 m) of significant wave height in deep-water (H_0) between the minimum and the maximum of the whole register were considered. For each H_0, three associated peak wave periods (T_p) were tested (Table 5.1): the minimum (T_{min}), the most frequent (T_{freq}), and the maximum (T_{max}).

5.2.2   Seabed Evolution with XBeach-G

The process-based model XBeach-G is an extension of the XBeach model that incorporates: (1) a non-hydrostatic pressure correction term that allows solving waves explicitly in model; (2) a groundwater model that allows infiltration and exfiltration;

Table 5.1 Sea states modelled with XBeach-G to simulate the morphological changes of the seabed in front of the reflective boundary

XBeach-G seabed profile	H_0 (m)	T_{min} (s) [A]	T_{freq} (s) [B]	T_{max} (s) [C]
P1	0.5	2	4	13.2
P2	1	2.6	4.8	13.3
P3	1.5	4.2	5.8	11.8
P4	2	5.2	6.9	11.9
P5	2.5	6.4	7.6	12.2
P6	3	7.1	7.7	10.7
P7	3.5	7.7	8.4	10.7
P8	4	8.4	8.7	9.6
P9	4.5	9.6	9.6	9.7
P10	5	9.6	9.6	9.6

and (3) the computation of bed load transport, including the effects of groundwater ventilation and flow acceleration forces, for estimating bed level changes [12, 13].

A 120 m–long flat bottom with a vertical boundary at the landward side, was simulated as the initial profile for each case. The input wave conditions were obtained from the Delft3D-WAVE model, which allowed the propagation of the deep-water sea states (Table 5.1) towards the nearshore. In addition, the Delft3D-model was calibrated through comparison with data collected by two Acoustic Doppler Current Profilers [2]. Values of sediment friction factor and *Nielsen*'s boundary layer phase lag used for the simulations were 0.03 and 20°, respectively, which were found to be optimum during the calibration of the XBeach-G for the study site [4]. The reflective boundary located at the landward limit of the computational domain were set as a non-erodible object. The final bed levels obtained with XBeach-G for each test case were used as input for the FLUENT® model, as detailed in the following section.

5.2.3 Wave Generation and OWC Simulation in FLUENT®

The bed morphology previously generated with XBeach-G were upgraded into a finite element domain to be used in a 2D flume designed in FLUENT®, including the OWC structure adjacent to the vertical breakwater. The flume has a length of 120 m and a height of 12 m, with an initial water depth of 5 m, according to those presented in [15]. The OWC was located at the right end of the mesh, with the Wells turbine represented by means of an Actuator Disk Model (ADM). The ADM was set as detailed in [15]. The paddle motion was accomplished by a dynamic mesh scheme. This model was validated through comparison with experimental data as shown in [14]. Time-dependent variables determining the instantaneous paddle position and velocity were implemented in a compiled UDF (User Defined Function) in the model, following the Bisel theory for a piston–type paddle [11]:

$$
\begin{aligned}
x_{paddle} &= \frac{S_0}{2} \cos \frac{2\pi t}{T} \\
U_{paddle} &= -\frac{S_0 \pi}{T} \sin \frac{2\pi t}{T} \\
S_0 &= H \frac{2k_0 h + \sinh 2k_0 h}{4 \sinh^2 k_0 h}
\end{aligned}
\tag{5.1}
$$

where x_{paddle} is the paddle displacement along the horizontal axis, U_{paddle} is the paddle displacement velocity, S_0 is the paddle stroke, H is the wave height, k is the wave number, h is the water depth at the paddle location, T is the wave period and t is the simulation time. The dynamic mesh ensures the harmonic oscillation of the free surface described by:

$$h_{owc} = h_{ref} - \eta$$
$$V_{owc} = S_{owc}(h_{ref} - h_{owc})$$
$$U_{owc} = \frac{d\eta}{dt} = -\frac{H\pi}{T}\sin\frac{2\pi t}{T}$$

(5.2)

where h_{owc} is the water surface level inside the OWC, h_{ref} is the reference or initial water level, η is the surface elevation, V_{owc} is the OWC chamber volume, S_{owc} is the OWC horizontal surface, and U_{owc} is the OWC water displacement velocity. The turbine performance is represented by means of the actuator disk model. The porous medium involved in the ADM definition is modelled by the addition of a momentum source term to the standard fluid flow equations. The source term consists of two parts: a viscous loss term (the first term on the right-hand side of Eq. (5.3)), and an inertial loss term (the second term on the right-hand side of Eq. (5.3) [10]:

$$\delta S_i = -\delta \left(\sum_{j=1}^{3} D_{ij}\mu v_j + \sum_{j=1}^{3} C_{ij}\frac{1}{2}\rho|v|v_j \right)$$

(5.3)

where S_i is the source term for the momentum equation, δ is the porous layer thickness, $|v|$ is the magnitude of the velocity, and D and C are prescribed matrices. This momentum source contributes to the pressure gradient in the porous cell, creating a pressure drop that is proportional to the fluid velocity (or velocity squared) in the cell. The units for the momentum source are N/m^3, so this term has to be multiplied by the porous layer thickness to insert it as a component of the external body forces in the momentum conservation equations [10]. Equation (5.3) can be adapted for a simple homogeneous porous media:

$$S_i = -\left(\frac{\mu}{\alpha}v_i + C_2\frac{1}{2}\rho|v|v_i \right)$$

(5.4)

where α is the permeability and C_2 is the inertial resistance factor.

The porous zone settings are the same for air and water phases, accounting for the fact that water never reaches the porous zone for the generated waves. The inertial resistance coefficient is set to $C_2 = 0$ for both x (horizontal) and y (vertical) directions. The viscous resistance coefficient is set to $1/\alpha = 10^7$ m^{-2} in the y direction, while for the x direction is set to $1/\alpha = 10^{10}$ m^{-2} in order to prevent undesired radial flow effects. The viscous resistance coefficients are adjusted following the measurements for a Wells turbine in a wind tunnel [15]. The adjustment of coefficients ensures the same relationship between pressure drop and flow velocity as in the experiments.

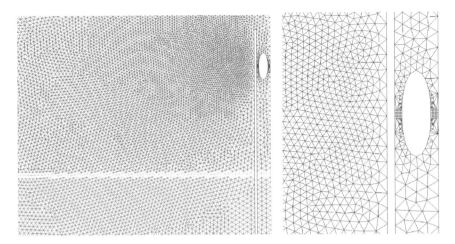

Fig. 5.1 Detailed view of OWC device in the numerical flume (left panel) and mesh detail around the porous layer (right panel). Mesh zone in blue is air, red zone is water, inter-phase in colour gradient. (*Source* [14]. Reproduced with permission of Elsevier)

5.2.4 FLUENT® Model Set Up

Test cases in FLUENT® were set under the same wave conditions used to generate bedforms with XBeach-G. Two conditions were considered for each test case: waves propagating over the flat bed and wave action over the evolved sea bed. This way, the effects of changes in bed level on the OWC performance were analysed.

The numerical domain was meshed in GAMBIT® under a Tri/Pave scheme with maximum spacing setting of 0.1 m and a minimum element size of 0.001 m (Fig. 5.1). A pre–meshing scheme was applied to the hub edges inside the chamber, with spacing of 0.01 m to help obtaining a smoother structure in the final mesh. A mesh with 343420 elements was generated. The FLUENT® solver was configured to laminar and VOF (Volume Of Fluid) with air as phase 1 and water as phase 2. The pressure–velocity coupling was set to PISO (Pressure Implicit Split Operator) and the discretionary scheme for pressure was set to PRESTO (Pressure Staggering Options), while for momentum equations was set to first order upwind. The free surface reconstruction was set to geo-reconstruct.

A set of six wave gauges were arranged in the numerical flume, as specified in Fig. 5.2. Records of total and static pressure, horizontal and vertical velocity, and density distribution, were obtained with a sampling period of 0.1 s.

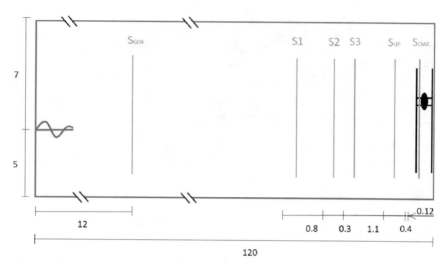

Fig. 5.2 Schema of gauges located in the numerical flume. Units in metres. (*Source* [14]. Reproduced with permission of Elsevier)

5.3   Results

5.3.1   Seabed Forms

Table 5.2 shows the mean amplitude ($\bar{\delta}$) and length ($\bar{\lambda}$) of the bedform with the highest unit volume (calculated by the trapezoidal rule) in each case. It is observed that the bedforms dimensions are generally related to the combination of wave height and period (wavelength), so that greater values of wave height and period generate higher bed level changes. The mean amplitudes appear to be directly influenced by the wave height; whereas they increase with the wave period (wavelength) up to values of around 9.6 s (145 m) and decrease for higher values. The bedform lengths are longer for higher wave heights and periods (wavelengths), but they seem to be more related to wave period (wavelength) since an almost linear trend is observed between the bedforms length and the wave period (wavelength).

These patterns are well observed in Fig. 5.3, which depicts the bed level in the proximity of the reflective boundary for eight representative cases selected to study the influence of seabed morphology on the OWC performance. The higher bedform amplitude (0.471 m) is generated by the case 10B, which is the one with highest wave height (5 m), but not with the highest wave period and wavelength (9.6 s and 143.9 m, respectively). On the other hand, longer bedforms lengths are obtained for greater values of wave period and length in most of the cases, with the only exception being the case P2C. Values of lengths and amplitudes of the generated bedforms for these cases are in the ranges of (0.159 m, 0.471 m) and (22.7 m, 42.3 m), respectively.

Table 5.2 Bedform mean amplitude ($\bar{\delta}$) and mean wavelength ($\bar{\lambda}$) for each case

Seabed profile	H (m)	T (s)	L (m)	$\bar{\delta}$ (m)	$\bar{\lambda}$ (m)
P1A	0.5	2	6.25	–	–
P1B	0.5	4	24.98	–	–
P1C	0.5	13.2	272.04	0.004	25
P2A	1	2.6	10.55	–	–
P2B	1	4.8	35.97	–	–
P2C	1	13.3	276.18	0.258	27.5
P3A	1.5	4.2	27.54	–	–
P3B	1.5	5.8	52.52	0.003	12.4
P3C	1.5	11.8	217.4	0.279	42.3
P4A	2	5.2	42.22	0.004	19.9
P4B	2	6.9	74.33	0.159	22.7
P4C	2	11.9	221.1	0.294	42.5
P5A	2.5	6.4	63.95	0.132	22.4
P5B	2.5	7.6	90.18	0.356	25.1
P5C	2.5	12.2	232.39	0.308	30.2
P6A	3	7.1	78.71	0.287	22.6
P6B	3	7.7	92.57	0.368	24.9
P6C	3	10.7	178.75	0.409	32.4
P7A	3.5	7.7	92.57	0.387	25.1
P7B	3.5	8.4	110.17	0.379	27.5
P7C	3.5	10.7	178.75	0.396	35
P8A	4	8.4	110.17	0.377	29.9
P8B	4	8.7	118.18	0.442	27.6
P8C	4	9.6	143.89	0.386	29.8
P9A	4.5	9.6	143.89	0.425	32.5
P9B	4.5	9.6	143.89	0.425	32.5
P9C	4.5	9.7	146.9	0.425	32.6
P10A	5	9.6	143.89	0.471	34.9
P10B	5	9.6	143.89	0.471	34.9
P10C	5	9.6	143.89	0.471	34.9

The influence of these forms on the OWC performance is detailed in the following sections.

5.3.2 Water Surface Elevation

Figure 5.4 depicts the surface elevation inside the OWC and at the generation zone for eight test cases, both flat bottom and evolved seabed. There are similarities for

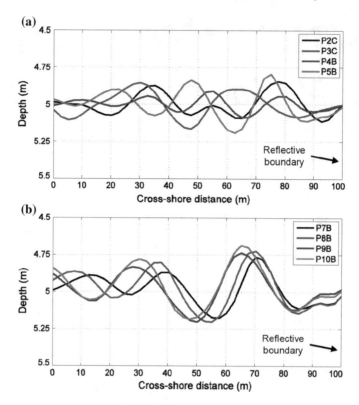

Fig. 5.3 Final bed levels obtained with the XBeach-G model: **a** Cases P2C, P3C, P4B and P5B, **b** P7B, P8B, P9B and P10B. (*Source* [14]. Reproduced with permission of Elsevier)

the smaller wave heights, i.e. tests P2C to P5B, between flat bottom and evolved seabed cases, both for generated waves and the waves inside the OWC. Conversely, for tests P7B to P10B, significant differences in the surface elevation inside the OWC are observed between flat bottom and changed seabed. The amplitude of the surface elevation is different in both cases, eventually showing a slight phase shift. This analysis shows that the changes in the bedforms affect the amplitude and the phase of the water surface inside the OWC. The larger the wave height, the more significant effect on the surface elevation.

Moreover, Fig. 5.5 shows the hydrodynamic efficiency in energy extraction (ϵ) as presented in [17]. The efficiency is calculated as the ratio between the pneumatic power available to the turbine (W) and the wave energy flux (f_E) averaged in a wave cycle (W/f_E). From the standpoint of bed evolution, higher efficiency values are observed for the smoothest bedforms, i.e. tests with intermediate kh values such as P3C ($H = 1.5$ m, $T = 11.8$ s), where h is the initial wave depth. In addition, when comparing for each single case, efficiency values are, in general terms, lower for the case with evolved seabed, exhibiting more noticeable differences for lower values of

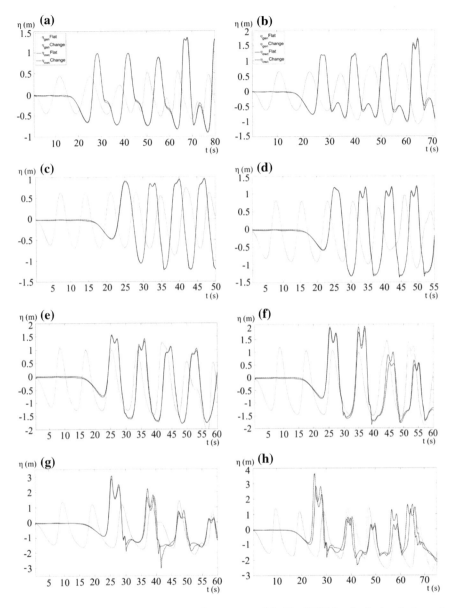

Fig. 5.4 Surface elevation in the generation gauge and inside OWC for flat bed and changed seabed. **a** P2C: $H = 1$ m, $T = 13.3$ s; **b** P3C: $H = 1.5$ m, $T = 11.8$ s; **c** P4B: $H = 2$ m, $T = 6.9$ s; **d** P5B: $H = 2.5$ m, $T = 7.6$ s; **e** P7B: $H = 3.5$ m, $T = 8.4$ s; **f** P8B: $H = 4$ m, $T = 8.7$ s; **g** P9B: $H = 4.5$ m, $T = 9.6$ s; **h** P10B: $H = 5$ m, $T = 9.6$ s. (*Source* [14]. Reproduced with permission of Elsevier)

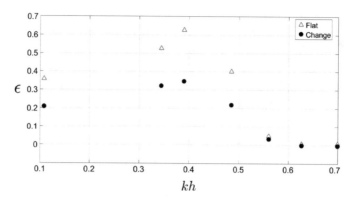

Fig. 5.5 Efficiency of energy extraction versus kh. (*Source* [14]. Reproduced with permission of Elsevier)

kh (up to $kh = 0.5$), e.g., wave periods over 8 s. The mean efficiency in the flat case is around 19%, while for the evolved seabed cases the mean efficiency is near 15%. This values match with those presented by [17], where the hydrodynamic efficiencies were close to 18% when an optimal phase control was applied. The present results show that for a real bedform those efficiency values decrease more than 4%.

It has to be noticed, however, that the generated wave slightly changes for the last cases (P10B, P9B and P8B). This effect might be caused by the wave height value, along with the possible reflection at the leeward end of the flume. High waves combined with moderate depths, in fact diminished by seabed forms, tend to shoal. This leads to an increase in crest height with respect to wave trough. This fact can be better understood by calculating the *Ursell number*, which determines the relation between wave non–linearity as described by H/L and wave dispersion:

$$U = \frac{H}{L}\left(\frac{L}{h}\right)^2 \tag{5.5}$$

where L is the wavelength. The *Ursell number* for the eight cases studied is depicted in Fig. 5.6. Case P8B shows the highest U, followed by P9B and P10B. This fact verifies the hypothesis of the wave slope affecting the wave pattern in different cases.

Figure 5.7 show the mean surface elevation and mean wave height, respectively, non-dimensionalized by the bedform mean amplitude ($\bar{\delta}$) shown in Table 5.2. The variations in the water surface oscillation inside the OWC following different changes in the seabed depends on the wavelength of the generated wave. As shown in Fig. 5.7, the surface elevation related to the bedform amplitude is higher in the case with changes in the seabed (black dots are placed over grey triangles in Fig. 5.7). There is a clear pattern as a function of kh, with marked trends for $kh \in [0.1, 0.4]$ and $kh \in [0.4, 0.7]$. In the following section, implications related to wave energy extraction are analysed.

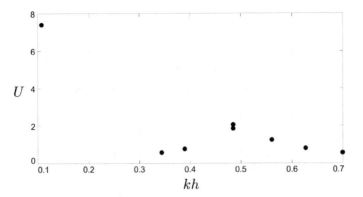

Fig. 5.6 Ursell number U versus kh. (*Source* [14]. Reproduced with permission of Elsevier)

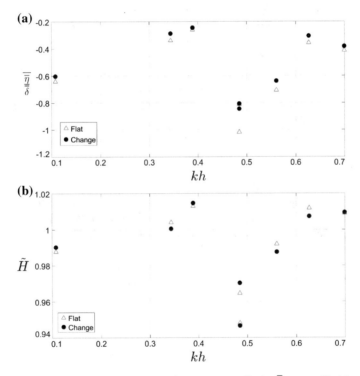

Fig. 5.7 Mean surface elevation ($\bar{\eta}$) over bedform mean amplitude ($\bar{\delta}$) versus kh (**a**), and mean wave height inside OWC over generated wave height ($\tilde{H}$) versus kh (**b**). (*Source* [14]. Reproduced with permission of Elsevier)

5.3.3 Wave Spectra

The generated spectra (and hence the energy) is mostly equal in every case, for the flat and evolved seabed (Fig. 5.8). The available energy inside the OWC is presented in a non-dimensional way, related to the flat case. From case P5B onwards, there is residual energy above $f = 0.5$ Hz present in the flat case. That means that the energy left inside the OWC for the flat case is higher than in the changed seabed case. For lower wave heights (first four cases, P2C to P5B), there is a constructive superposition of the energy inside the OWC related to the energy generated; whereas for bigger waves (last four cases, P7B to P10B), the peak energy inside the OWC is smaller than the peak in the generated spectra, but the shape of the spectrum is also more chaotic. For higher waves there is an extra energy in the flat cases. Moreover, peaks in the spectra inside the OWC (right column in Figs. 5.8 and 5.9) show different positions for the cases with evolved sea bed with respect to the flat cases. This fact highlights the change in the absorption patterns due to seabed forms.

This phenomena could be explained because of the bedforms: these changes in the seabed imply more friction at the flume bottom, which extracts energy from the wave flow. This energy dissipated by the bedform is then extracted from the flow, resulting in a lower amount of energy arriving to the OWC. Moreover, changes in the streamlines form around the bedforms, previously straight and parallel to the flume bottom, produce eddies that modify the flow dynamics.

Figure 5.10 shows the velocity contours and velocity vectors (white dots) for the case P7B. Although the velocity magnitude is quite similar in both cases, the velocity vectors near the bottom flume are very different. In the flat case, vectors are parallel to the bottom. In the case with bedforms, velocity vectors near the bottom are sinusoidal, showing a pattern opposite to the shape of the bedform, e.g. when the bedform is concave, the streamline over it is convex. In Fig. 5.10, velocity vectors coloured by stream function are presented. In this case, despite the similitude in the velocity field between flat and changed seabed cases, the mass transport along the flume in the case with bedforms is nearly twice than in the flat case. Moreover, the transport geometrical patterns are different. This fact, linked to the shape of the streamlines, is an indicative of the change in the energy distribution generated by the bedforms.

5.3.4 Pressure Drop Inside OWC Versus Air Flow

In Fig. 5.11, case P3C has a very representative airflow amplitude in comparison with its wave height (1.5 m). This case corresponds with $kh = 0.39$, which is the case with the largest bedform wavelength. The larger the bedform wavelength, the bigger the airflow inside the OWC. Cases P10B and P7B also have a very representative airflow, which correspond to $kh = 0.49$ and $kh = 0.56$, respectively. The maximum in the bedform amplitude is located in these points. The larger the bedform amplitude, the bigger the airflow inside the OWC. On the other hand, case P8B presents a bedform

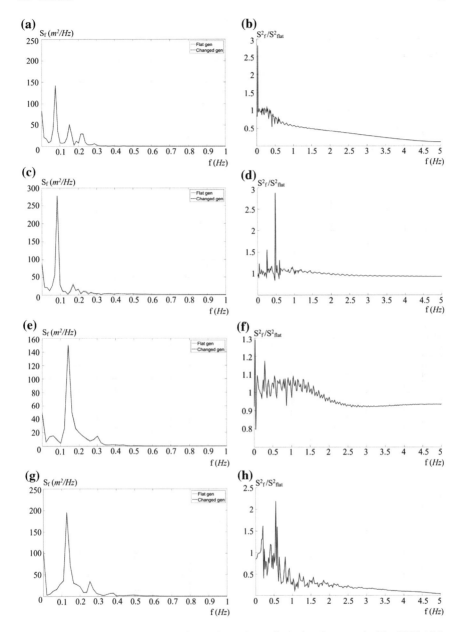

Fig. 5.8 Generated wave spectra (left panels) and non-dimensional spectra inside OWC (right panels) for flat bed and changed seabed: **a** and **b**: P2C ($H = 1$ m, $T = 13.3$ s); **c** and **d**: P3C ($H = 1.5$ m, $T = 11.8$ s); **e** and **f**: P4B ($H = 2$ m, $T = 6.9$ s); **g** and **h**: P5B ($H = 2.5$ m, $T = 7.6$ m). (*Source* [14]. Reproduced with permission of Elsevier)

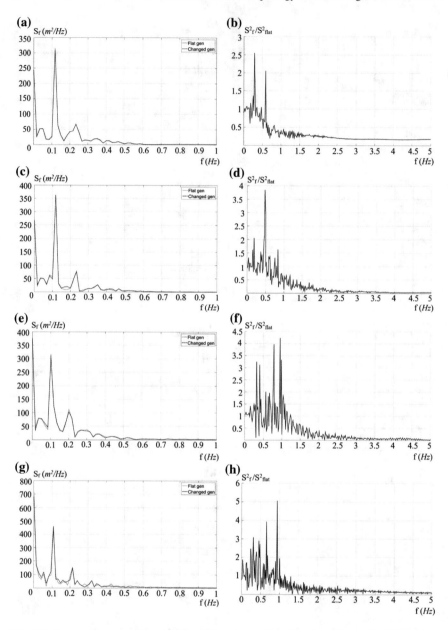

Fig. 5.9 Generated wave spectra (left panels) and non-dimensional spectra inside OWC (right panels) for flat bed and changed seabed: **a** and **b**: P7B ($H = 3.5$ m, $T = 8.4$ s); **c** and **d**: P8B ($H = 4$ m, $T = 8.7$ s); **e** and **f**: P9B ($H = 4.5$ m, $T = 9.6$ s); **g** and **h**: P10B ($H = 5$ m, $T = 9.6$ s). (*Source* [14]. Reproduced with permission of Elsevier)

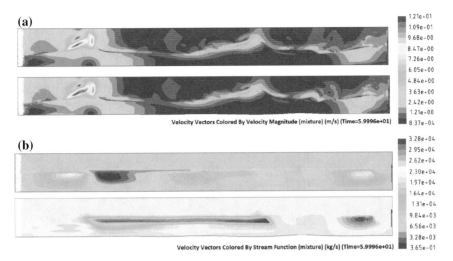

Fig. 5.10 Velocity vectors coloured by velocity magnitude (m/s) (**a**) and stream function (kg/s) (**b**). Flat bottom (upper panels) and evolved seabed (bottom panel) for the case P7B at the end of the simulation time ($t = 60\,s$). (*Source* [14]. Reproduced with permission of Elsevier)

amplitude of 0.442 m, one of the highest of the presented cases, but a bedform wavelength of 27.6 m, which gives a moderate airflow amplitude. Finally, case P4B shows both moderate bedform amplitude and wavelength, and presents a limited pressure drop–airflow amplitude, very close to the values shown for case P2C, even when the wave height is twice in the first case. Then, the bedforms affect the pressure drop–airflow relationship inside the OWC, e.g., they change the slope of the $\Delta p - Q$ curve and the range. Moreover, the maximum change in the conditions inside the OWC depends on both the maximum bedform wavelength and maximum bedform amplitude, which implies that these changes depend on the bedform unit volume, instead of on single geometrical parameters.

5.4 Conclusions

This chapter presents a numerical study on the effect of seabed changes on OWC devices performance. A series of tests were run in XBeach-G to reproduce the generation of bedforms in front of reflective boundaries based on wave conditions of a gravel-dominated beach (Playa Granada, southern Spain). These seabed morphologies were used as input by the FLUENT model to analyse the effect of bedforms on an OWC compared to a flat seabed ideal case. Based on the results, the following conclusions are drawn:

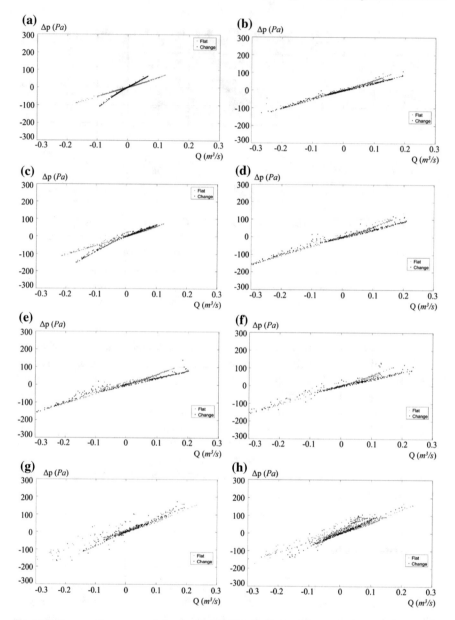

Fig. 5.11 Pressure drop versus air flow inside OWC for flat bed and changed seabed. **a**: P2C ($H = 1$ m, $T = 13.3$ s); **b**: P3C ($H = 1.5$ m, $T = 11.8$ s); **c**: P4B ($H = 2$ m, $T = 6.9$ s); **d**: P5B ($H = 2.5$ m, $T = 7.6$ m); **e**: P7B ($H = 3.5$ m, $T = 8.4$ s); **f**: P8B ($H = 4$ m, $T = 8.7$ s); **g**: P9B ($H = 4.5$ m, $T = 9.6$ s); **h**: P10B ($H = 5$ m, $T = 9.6$ s). (*Source* [14]. Reproduced with permission of Elsevier)

- The dimensions of the bedforms generated in front of reflective boundaries have been demonstrated to be dependent on the combination of wave height and period. The bedform amplitudes are more influenced by the wave height; whereas the bedform lengths are more related to the wave period. Thus, by means of wave climate data, bedforms dimensions can be predicted, with direct implications on the OWC performance.

- The highest hydrodynamic efficiency, calculated according to [17], is attained in tests with intermediate kh values, such as P3C ($H = 1.5$ m, $T = 11.8$ s). In any case efficiency values are lower for the case with evolved seabed, showing more noticeable differences for wave periods over 8 s. The flat cases present a mean efficiency (19%) similar to that presented by [17] (18%), while mean efficiency for the evolved bedform cases decreases up to 15%.

- The mean surface elevation over the mean bedform amplitude ($\overline{\eta}/\overline{\delta}$) is, in general, slightly larger for the cases with bedforms. This fact can be explained with the *Ursell* number (U). It was calculated to study the non-linear effects in waves. The greater values of U were obtained for the three cases with the highest wave heights (P8B, P9B and P10B). This indicates that the wave slope is being affected by the bedform and is changing the surface elevation pattern for these cases.

- Although the wave velocity magnitude is the same for the flat and seabed forms cases, the streamlines change their forms. While in the flat cases the streamlines are flat and parallel to the surface, in the seabed forms cases they show a radial pattern centred in the forms troughs. Despite the lack of change in the velocity magnitude, the mass transport along the flume in the changed seabed cases is nearly twice the mass transport in the flat case. Also, the transport patterns change, showing a more extended transport along the flume direction. These factors imply that a change in the energy distribution inside the flume is caused by the bedforms.

- The generated wave spectra is the same for flat and changed seabed cases. However, the wave spectra inside the OWC changes in shape and value. Particularly, there is a residual energy left in the flat case that is not present in the seabed form case. Moreover, changes in the position of peak energy periods inside the OWC are observed for the cases with evolved seabed in comparison with the flat cases.

- The bedforms unit volume directly affects the OWC performance. The slope of the pressure drop–air flow curve for the OWC turbine changes depending on the unit volume. However, this result might not be conclusive: pressure drop is smaller for higher bedform unit volume, but air flow grows. Thus, the power output may not change in value, but the way this power is absorbed is changed. This fact, linked to the residual energy in the wave spectra of the flat cases make us conclude that there is a energy dissipation induced by the seabed that is reflected in a lack of absorption in the OWC.

- Changes in the $\Delta p - Q$ curve slope influence the absorption patterns of the device depending on the location and wave conditions. Therefore, the power matrix of a device will change depending on the changes in the seabed.

This chapter shows that bedforms dimensions have a strong influence in water flow circulation patterns. Those directly influence OWC performance by modifying the incoming flow, and thus changing the pressure–air flow behaviour in the turbine. Local seabed conditions around devices should be studied in future plans for OWC potential locations in order to accurately estimate the device performance and efficiency.

Acknowledgements The second author was funded by the TALENTIA Fellowship Programme (Regional Ministry of Economy, Innovation, Science and Employment, Andalusia, Spain) and the third author was funded by the Research Contract BES-2013-062617 (Spanish Ministry of Economy and Competitiveness).

References

1. Bergillos, R.J., López-Ruiz, A., Medina-López, E., Moñino, A., Ortega-Sánchez, M.: The role of wave energy converter farms on coastal protection in eroding deltas, Guadalfeo, southern Spain. J. Clean. Prod. **171**, 356–367 (2018)
2. Bergillos, R.J., López-Ruiz, A., Ortega-Sánchez, M., Masselink, G., Losada, M.A.: Implications of delta retreat on wave propagation and longshore sediment transport-Guadalfeo case study (southern Spain). Mar. Geol. **382**, 1–16 (2016)
3. Bergillos, R.J., López-Ruiz, A., Principal-Gómez, D., Ortega-Sánchez, M.: An integrated methodology to forecast the efficiency of nourishment strategies in eroding deltas. Sci. Total Environ. **613**, 1175–1184 (2018)
4. Bergillos, R.J., Masselink, G., McCall, R.T., Ortega-Sánchez, M.: Modelling overwash vulnerability along mixed sand-gravel coasts with XBeach-G: Case study of Playa Granada, southern Spain. In: Coastal Engineering Proceedings, vol. 1, Issue 35, p. 13 (2016)
5. Bergillos, R.J., Masselink, G., Ortega-Sánchez, M.: Coupling cross-shore and longshore sediment transport to model storm response along a mixed sand-gravel coast under varying wave directions. Coast. Eng. **129**, 93–104 (2017)
6. Bergillos, R.J., Ortega-Sánchez, M.: Assessing and mitigating the landscape effects of river damming on the Guadalfeo River delta, southern Spain. Landsc. Urban Plan. **165**, 117–129 (2017)
7. Bergillos, R.J., Ortega-Sánchez, M., Masselink, G., Losada, M.A.: Morpho-sedimentary dynamics of a micro-tidal mixed sand and gravel beach, Playa Granada, southern Spain. Mar. Geol. **379**, 28–38 (2016)
8. Bergillos, R.J., Rodríguez-Delgado, C., Millares, A., Ortega-Sánchez, M., Losada, M.A.: Impact of river regulation on a Mediterranean delta: Assessment of managed versus unmanaged scenarios. Water Resour. Res. **52**, 5132–5148 (2016)
9. Bergillos, R.J., Rodríguez-Delgado, C., Ortega-Sánchez, M.: Advances in management tools for modeling artificial nourishments in mixed beaches. J. Mar. Syst. **172**, 1–13 (2017)
10. Fluent, I.: FLUENT 6.3 Users guide-7.19.6 user inputs for porous media. Fluent Inc., Centerra Resource Park, 10 Cavendish Court, Lebanon, NH 03766 (2006)
11. Hughes, S.A.: Physical models and laboratory techniques in coastal engineering, vol. 7. World Scientific (1993)
12. McCall, R.T., Masselink, G., Poate, T.G., Roelvink, J.A., Almeida, L.P.: Modelling the morphodynamics of gravel beaches during storms with XBeach-G. Coast. Eng. **103**, 52–66 (2015)
13. McCall, R.T., Masselink, G., Poate, T.G., Roelvink, J.A., Almeida, L.P., Davidson, M., Russell, P.E.: Modelling storm hydrodynamics on gravel beaches with XBeach-G. Coast. Eng. **91**, 231–250 (2014)

14. Medina-López, E., Bergillos, R., Moñino, A., Clavero, M., Ortega-Sánchez, M.: Effects of seabed morphology on oscillating water column wave energy converters. Energy **135**, 659–673 (2017)
15. Moñino, A., Medina-López, E., Clavero, M., Benslimane, S.: Numerical simulation of a simple OWC problem for turbine performance. Int. J. Mar. Energy **20**, 17–32 (2017)
16. Sánchez-Badorrey, E., Losada, M., Rodero, J.: Sediment transport patterns in front of reflective structures under wind wave-dominated conditions. Coast. Eng. **55**(7–8), 685–700 (2008)
17. Sarmento, A., Gato, L., Falcão, AdO: Turbine-controlled wave energy absorption by oscillating water column devices. Ocean Eng. **17**(5), 481–497 (1990)

Chapter 6
The Role of Wave Energy Converter Farms in Coastal Protection

Abstract Many worldwide coasts are under erosion with climate projections indicating that damages will rise in future decades. Specifically, deltaic coasts are highly vulnerable systems due to their low-lying characteristics. This chapter investigates the role of wave energy converter (WEC) farms on the protection of an eroding gravel-dominated deltaic coast (Guadalfeo, southern Spain). Eight scenarios with different alongshore locations of the wave farm were defined and results were compared with the present (no farm) configuration of the coast. Assuming that storm conditions drive the main destruction to the coast, we analysed the impact of the most energetic storm conditions and quantified the effects of the location of the farm. Significant wave heights in the lee of the farm were calculated by means of a calibrated wave propagation model (Delft3D-Wave); whereas wave run-up and morphological changes in eight beach profiles were quantified by means of a calibrated morphodynamic model (XBeach-G). The farm induces average reductions in significant wave heights at 10 m water depth and wave run-up on the coast down to 18.3% and 10.6%, respectively, in the stretch of beach most affected by erosion problems (Playa Granada). Furthermore, the erosion of the beach reduces by 44.5% in Playa Granada and 23.3% in the entire deltaic coast. Combining these results with previous works at the study site allowed selecting the best alternative of wave farm location based not only on coastal protection but also on energetic performance criteria. This chapter, whose methodology is feasibly extensible to other coasts worldwide, provides insights into the role of the alongshore location of WEC farms on wave propagation, run-up and morphological storm response of deltaic coasts.

6.1 Objective

The main objective of this chapter is to quantify and analyse the influence of the alongshore location of wave farms on the hydro- and morphodynamics of a gravel-dominated deltaic coast (Guadalfeo, southern Spain) under storm conditions. To meet these goals, wave propagation patterns, wave run-up and morphological changes of the beach were assessed by means of the application of calibrated wave propagation and morphodynamic models to eight different scenarios of wave farms location.

© The Author(s) 2018
A. Moñino et al., *Thermodynamics and Morphodynamics in Wave Energy*,
SpringerBriefs in Energy, https://doi.org/10.1007/978-3-319-90701-7_6

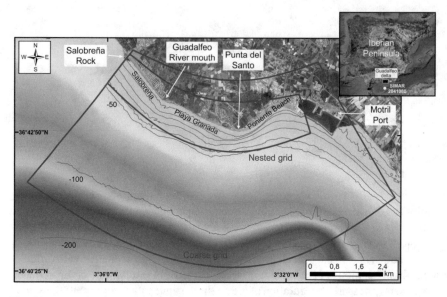

Fig. 6.1 Upper right panel: Location of the study site (Guadalfeo delta, southern Spain). Central panel: bathymetric contours, grids used in the wave propagation model, and location of the Salobreña Rock, Guadalfeo River mouth, Playa Granada, Punta del Santo, Poniente beach and Motril Port. (*Source* [5]. Reproduced with permission of Elsevier)

All results were compared to the current situation (no farm) and the best alternative was selected, based on energetic performance and coastal protection criteria. The latter was quantified as the reductions in nearshore wave height, wave run-up and beach erosion.

6.2   Study Site

The Guadalfeo deltaic coast is a 6.8-km-long micro-tidal beach located on the southern coast of Spain that faces the Mediterranean Sea (Fig. 6.1). It is bounded to the west by Salobreña Rock and to the east by Motril Port [18]. The Guadalfeo River contributes most sediment to the beach [14]. Its basin covers an area of 1252 km^2, including the highest peaks on the Iberian Peninsula ($\sim$3400 m.a.s.l.), and the river is associated with one of the most high-energy drainage systems along the Spanish Mediterranean coast [36].

The river was dammed 19 km upstream from the mouth in 2004, regulating 85% of the basin run-off [27]. As a consequence of river damming, the delta currently experiences coastline retreat, severe erosion problems [13] and frequent coastal flooding (Fig. 6.2). The central stretch of beach (Playa Granada) has been particularly affected and has been subjected to higher levels of coastline retreat in recent years than both western and eastern stretches, known as Salobreña and Poniente Beach,

Fig. 6.2 **a** Storm-induced erosion problems in Playa Granada (March 2014), **b** coastal flooding in Poniente Beach (February 2017). (*Source* [5]. Reproduced with permission of Elsevier)

respectively (Fig. 6.1). Consequently, artificial nourishment projects have been frequently conducted since the entry into operation of the dam [11]. However, the success of these interventions has been very limited since they lasted on average less than three months [9].

The particle size distribution on the studied coastal area presents varying proportions of sand and gravel [10], with three predominant fractions: sand (~0.35 mm), fine gravel (~5 mm) and coarse gravel (~20 mm). However, the morphodynamic response of the beach is dominated by the coarse gravel fraction due to the selective removal of the finer material [12] and the reflective shape of the profile is similar to those found on gravel beaches [30, 38]. Previous numerical works also demonstrated that the best fits to the measured profiles [7] and shorelines [15] are obtained by assuming that the beach is made up of coarse gravel.

The region is subjected to the passage of extra-tropical Atlantic cyclones and Mediterranean storms, with average wind speeds of 18–22 m/s [37], which generate wind waves under fetch-limited conditions (approximately 200–300 km). The storm wave climate is bimodal with prevailing west-southwest (extra-tropical cyclones) and east-southeast (Mediterranean storms) wave directions [6]. The 90%, 99% and 99.9% not exceedance significant wave heights in deep water are 1.2 m, 2.1 m and 3.1 m, respectively. The astronomical tidal range is ~0.6 m, whereas typical storm surge levels can exceed 0.5 m [12].

6.3   Methodology

To evaluate the efficiency of a wave farm as a coastal defence, the impacts of extreme south-westerly and south-easterly storms ($H_{99.9\%}$) were simulated using the Delft3D-Wave model and an extension of the XBeach-G model along the entire deltaic coast for eight different scenarios (Sect. 6.3.1). Each scenario corresponds to a different alongshore location of the wave farm (Fig. 6.3). These models were used because they are able to reproduce wave propagation patterns and storm-driven morphological changes, and they have been calibrated for the study site (Sects. 6.3.2 and 6.3.3).

The modelled wave variables were $H_0 = 3.1$ m, $T_p = 8.4$ s (the most frequent period under storm conditions), $\theta_{0,SW} = 238°$ and $\theta_{0,SE} = 107°$ (the most frequent directions under south-westerly and south-easterly storms, respectively). These constant sea states, summarized in Table 6.1, were simulated considering a storm surge of 0.5 m for a duration of 6 hours around high tide (tidal peak of 0.3 m), according to [8]. These wave conditions were tested because storms are the main responsible of the erosion problems and coastal flooding in the study site.

6.3.1   Wave Farm Locations and Geometry: Scenarios

López-Ruiz et al. [24] quantified the wave energy resource in 24 locations along the Guadalfeo deltaic coast during a 25-year period (typical lifetime of the WECs according to [20, 28], among others). The highest mean and extreme values of wave power were obtained at 30 m water depth, being the reason to define the eight wave farm scenarios (Fig. 6.3) with locations centred at this depth (A30–H30 in [24]). This range of water depths for the wave farm location is also in agreement with previous works (e.g., [2, 35] or [25]).

Thus, we can compare and select the best scenario considering not only the availability of wave power, but also the hydrodynamic and morphological effects of the wave farm on the coast. Each farm contains eleven WaveCat WECs arranged in two rows, with a distance between the twin bows of a single WaveCat WEC equal to 90 m [16] and a distance between devices equal to 180 m (Fig. 6.3), following the geometrical layout proposed in previous works (e.g., [1, 3]). WaveCats are overtopping

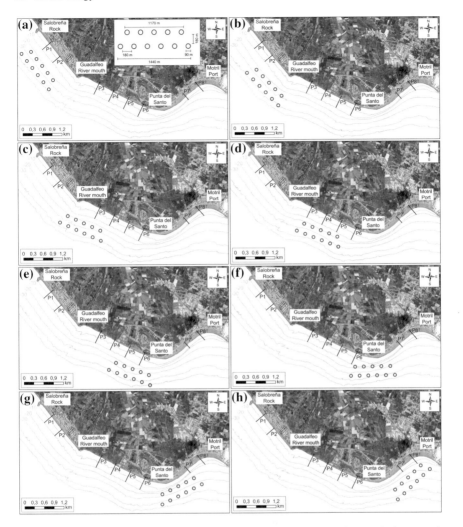

Fig. 6.3 Geometry and locations of the modelled wave farms: **a** scenario 1, **b** scenario 2, **c** scenario 3, **d** scenario 4, **e** scenario 5, **f** scenario 6, **g** scenario 7, **h** scenario 8. The study beach profiles are shown (P1–P8). (*Source* [5]. Reproduced with permission of Elsevier)

WECs, developed by [19], which are moored to the sea bottom by a single chain, allowing the self-orientation of the device with the local wave directions [4].

Each scenario was simulated by means of the wave propagation model detailed in Sect. 6.3.2 to evaluate the influence of the farm on the significant wave height. In addition, the wave run-up and morphological changes were obtained in eight selected beach profiles (P1–P8, Fig. 6.3) through an extension of the XBeach-G model (Sect. 6.3.3). To evaluate the reductions in wave height, run-up and beach erosion induced by the wave farm, the weighted averages among the number of storms

Table 6.1 Sea-states modelled with Delft3D-Wave and an extension of XBeach-G to study the influence of wave farm on the coastal morphodynamics for each scenario

	SW storm	SE storm
H_0 (m)	3.1	3.1
T_p (s)	8.4	8.4
θ_0 (°)	238	107
η_{ss}	0.5	0.5

($H_0 \geq 3.1$) incoming from each direction over last 25 years (lifetime period) were calculated for each scenario and the results were compared to the current configuration of the coast, which was noted as scenario 0.

6.3.2   Wave Propagation Model: Delft3D-Wave

The sea-states shown in Table 6.1 were propagated from deep-water to the nearshore using the WAVE module of the Delft3D model [22, 23], which is based on the SWAN model [21]. The main processes included in the model are refraction due to bottom and current variations; shoaling, blocking and reflections due to opposing currents; transmission/blockage through/by obstacles; wind effects; whitecapping; depth-induced wave breaking; bottom friction; and non-linear wave-wave interactions [26]. The results were used to address the significant wave heights in the leeward side of the farms and to provide the boundary conditions for the extension of the XBeach-G model for each scenario.

The model is able to simulate the effects of obstacles on wave propagation patterns, i.e., reduction of the wave height propagating behind or over the obstacle along its length, reflection of the waves that impinge the obstacle, and diffraction of the waves around its boundaries [41]. The WECs were included in the model as obstacles with circular shape, so that the devices always expose the same width to the incident waves, regardless the incoming wave direction, simulating the self-orientation behaviour of the device [25]. Based on the results of laboratory tests detailed by [19], we adopted constant mean values of the reflection and transmission coefficients equal to $k_r = 0.43$ and $k_t = 0.76$, respectively. These values are in agreement with those considered by [2] to model this type of WECs.

The model domain consisted of two different grids, shown in Fig. 6.1. The first is a coarse curvilinear 82×82-cell grid covering the entire deltaic region, with cell sizes that decrease with depth from 170×65 to 80×80 m. The second is a nested grid covering the beach area with 244 and 82 cells in the alongshore and cross-shore directions, respectively, and with cell sizes of approximately 20×15 m. This model was calibrated for these particular grids by [15] through comparison with field data, obtaining coefficients of determination higher than 0.86.

6.3.3   Morphodynamic Model: XBeach-G

The 1D process-based model XBeach-G is an extension of the XBeach model [39, 40] that incorporates: (1) a non-hydrostatic pressure correction term that allows solving waves explicitly in model; (2) a groundwater model that allows infiltration and exfiltration; and (3) the computation of bed load transport, including the effects of groundwater ventilation and flow acceleration forces, for estimating bed level changes [29, 32–34]. XBeach-G was combined with longshore sediment transport by [8] to make it more suitable for coasts highly influenced by both cross-shore and longshore sediment transport (like the study site). This coupled approach was applied to model the storm response of the study beach profiles (P1–P8, Fig. 6.3).

Measured topographic and bathymetric data were used as initial morphologies of the beach profiles. The input wave boundary conditions were obtained from the Delft3D-WAVE model at a depth of 10 m. This water depth offshore boundary fulfils all requirements detailed in the manual of the XBeach-G model [17], and is deeper than the closure depth in the study site [14, 15]. Values of sediment friction factor and Nielsen's boundary layer phase lag used for the simulations were 0.03 and 20°, respectively, which were found to be optimum during the calibration of the model [7, 8], demonstrating that it is able to reproduce the coastal processes under storm conditions.

The seawalls located landward of the profiles P1, P2, P7 and P8 (Fig. 6.3), and the infrastructure associated with the hotel complex located landward of the profiles P5 and P6 (Fig. 6.3) were included in the model as non-erodible and impermeable objects; whereas the farming settlements located landward of the profiles P3 and P4 were introduced as non-erodible and permeable boundaries. The results of the coupled model were used to quantify the wave run-up and the storm-driven morphological changes in the study profiles for each scenario.

6.4   Results

6.4.1   Wave Propagation

Figure 6.4 depicts the reductions in significant wave height for four scenarios of wave farm location with respect to scenario 0. Higher significant wave height reductions leewards of the farms are obtained under westerly storms (with values up to 1.4 m), whereas the area influenced by the farm is larger under easterly waves (Fig. 6.4a2–d2). This is particularly relevant in the stretch between Salobreña Rock and Punta del Santo due to the higher obliquity of easterly waves compared to westerly ones. As expected, the effects of scenarios 2, 4 and 8 are higher in Salobreña, Playa Granada and Poniente Beach, respectively; being almost negligible for the rest of beach stretches. On the other hand, the wave farm located in the vicinity of Punta del Santo (scenario 6) lead to wave propagation variations (compared to scenario

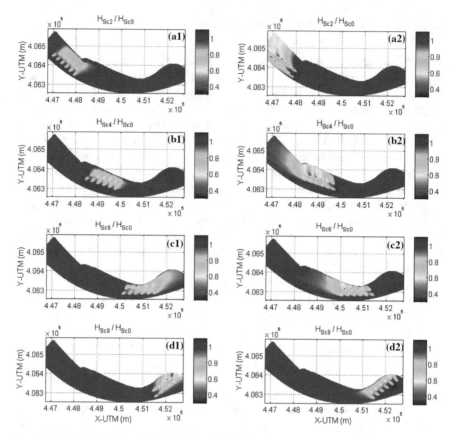

Fig. 6.4 Spatial distribution of the reductions in significant wave height for scenarios 2 (**a**), 4 (**b**), 6 (**c**), and 8 (**d**) with respect to scenario 0 under westerly (1) and easterly (2) storm conditions. (*Source* [5]. Reproduced with permission of Elsevier)

0) in both Playa Granada and Poniente Beach under easterly and westerly storms, respectively (Fig. 6.4c).

The significant wave heights at 10 m water depth (H_{10m}) are generally higher under westerly storm conditions (Fig. 6.5). The effects of the wave farm depending on its alongshore location are also clearly observed in Fig. 6.5, corresponding to the zone with oscillations in H_{10m}. If the values of westerly and easterly H_{10m} are averaged taking into account the number of storms incoming from each direction over last 25 years, the lowest H_{10m} both in the section subjected to a stronger erosion (Playa Granada, H_{PG}) and along the entire deltaic coast (H_T) are obtained for scenario 5, with values less than 1.74 m. Scenarios 3, 4 and 6 provide values of both H_{PG} and H_T lower than 2 m; whereas the highest H_{10m} are obtained for scenario 1 (Fig. 6.5).

The average reductions with respect to scenario 0 under westerly storm conditions are up to 7.7% along the entire deltaic coast and up to 15.7% in the section of

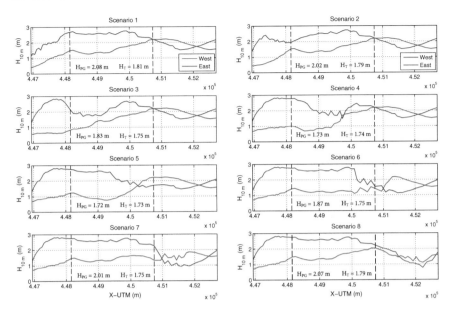

Fig. 6.5 Alongshore variation of the significant wave height at 10 m water depth for each scenario under westerly (blue) and easterly (red) storm conditions. The average values in Playa Granada (H_{PG}) and along the entire deltaic coast (H_T) are indicated. (*Source* [5]. Reproduced with permission of Elsevier)

Playa Granada (both for scenario 4). On the contrary, the maximum reductions under easterly storms are obtained for scenario 5, with average values of 24% in Playa Granada and 12.4% along the entire deltaic coast (Fig. 6.6). Considering again the weighted average of westerly and easterly storms, the maximum decreases in H_{10m} are reached for scenario 5, with reductions of 9.8% in the entire deltaic coast and 18.3% in Playa Granada (Table 6.2). This latter value is higher than those obtained on a sandy beach (Perranporth, UK) through wave farms at 30 m and 35 m water depth, with average reductions about 15% and 9%, respectively; but lower than that obtained by means of a wave farm at 25 m water depth (25%), according to [3].

Regarding the alongshore distances affected by wave farms, the highest protection under westerly waves is provided by scenario 6, with a total length of wave height reduction about 2300 m; whereas scenario 5 presents the highest influence under easterly storms, inducing wave height reductions along approximately 2400 m. In relative terms (i.e., altered length between total length), the greatest percentage (weighted average of westerly and easterly storms) along the entire deltaic coast is obtained with scenario 6. The highest protection in Playa Granada is provided by scenario 5, with more than 60% of the stretch affected by farm-induced decreases in significant wave height, as shown the bold values in Table 6.2.

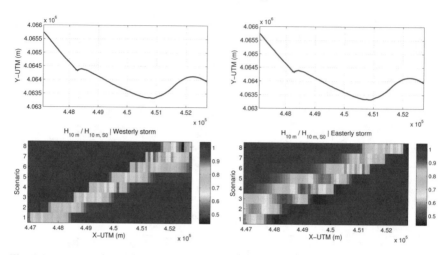

Fig. 6.6 Reductions in significant wave height at 10 m water depth for each scenario with respect to scenario 0 under westerly (left panel) and easterly (right panel) storm conditions. The shoreline is shown in the upper panels. (*Source* [5]. Reproduced with permission of Elsevier)

Table 6.2 Percent reduction in significant wave height at 10 m water depth (H_{10m}), wave run-up ($R_{2\%}$) and beach erosion (AV) for each scenario with respect to scenario 0. The percentages of alongshore distance protected by each scenario with respect to the coastline length (L) are also indicated

	H_{10m}		L		$R_{2\%}$		AV	
	Total (%)	PG (%)	Total (%)	PG (%)	Total (%)	PG (%)	Total (%)	PG (%)
Scenario 1	4.35	0.29	12.82	3.27	4.49	0.96	11.08	8.95
Scenario 2	5.69	2.96	17.14	7.78	4.17	1.83	2.62	4.44
Scenario 3	8.58	10.96	27.11	37.02	4.31	5.43	6.06	5.64
Scenario 4	9.27	16.99	31.45	54.97	4.87	8.56	**23.3**	**44.53**
Scenario 5	**9.84**	**18.31**	33.73	**61.9**	**7.73**	**10.57**	22.82	44
Scenario 6	8.77	11.94	**37.4**	40.58	5.32	8	14.67	19.15
Scenario 7	8.36	4.87	27.3	12.5	2.32	3.26	3.36	14.83
Scenario 8	6.84	0.74	22.59	0	3.03	1.14	5.09	4.34

6.4.2 Wave Run-Up

As reported in Sect. 6.2, coastal flooding is a common problem at the study site (Fig. 6.2b), being wave run-up the main contributor [7, 12]. For this reason, the 2% exceedence values of the wave run-up elevation above the tide and surge levels ($R_{2\%}$) were computed in the study beach profiles (P1–P8, Fig. 6.3) under both westerly and easterly storm conditions. This value of wave run-up ($R_{2\%}$) is typically used in coastal engineering applications (e.g., [42] or [31]). The results, shown in Fig. 6.7, highlight that $R_{2\%}$ is higher under westerly storms than under easterly ones. Considering both

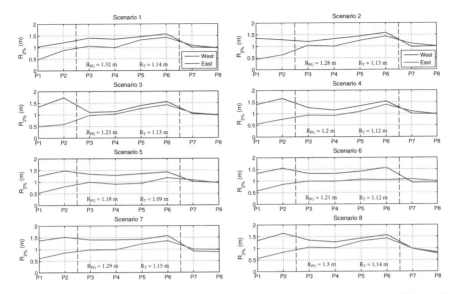

Fig. 6.7 Alongshore variation of the wave run-up for each scenario under westerly (blue) and easterly (red) storm conditions. The average values in the beach profiles of Playa Granada (R_{PG}) and in all beach profiles (R_T) are indicated. (*Source* [5]. Reproduced with permission of Elsevier)

directions, the minimum $R_{2\%}$ is reached with scenario 5: the average values are 1.09 m and 1.18 m along the entire deltaic coast and in Playa Granada, respectively.

Under westerly storms, the maximum reductions with respect to scenario 0 are equal to 7.8% along the entire deltaic coast (scenario 2) and 9.1% in Playa Granada (scenario 3); whereas under easterly storms the maximum decreases in wave run-up are obtained for scenario 5, with average values of 8.8% along the entire delta and 15.6% in Playa Granada (Fig. 6.8). Taking into account both westerly and easterly storms, the maximum reductions compared to scenario 0 are equal to 7.7% and 10.6% in the entire deltaic coast and in Playa Granada (both for scenario 5), respectively (Table 6.2).

6.4.3 Morphological Changes

To evaluate the effects of wave farms on beach morphology, the storm responses of the study beach profiles for each scenario were analysed and compared to the actual situation (scenario 0). All results reveal that profile responses are highly affected by the two or three scenarios with farms located in their proximity (leading to reductions in the erosion of the emerged profiles), with variations significantly lower for the rest of scenarios. These morphological changes are highly influenced by the incoming wave directions, partly due to the wave propagations patterns (Figs. 6.4 and 6.5).

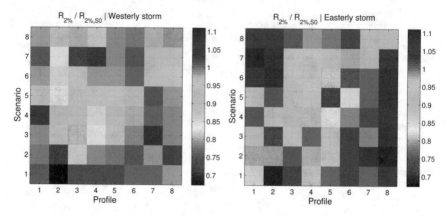

Fig. 6.8 Reductions in wave run-up for each scenario with respect to scenario 0 under westerly (left panel) and easterly (right panel) storm conditions in profiles P1–P8. (*Source* [5]. Reproduced with permission of Elsevier)

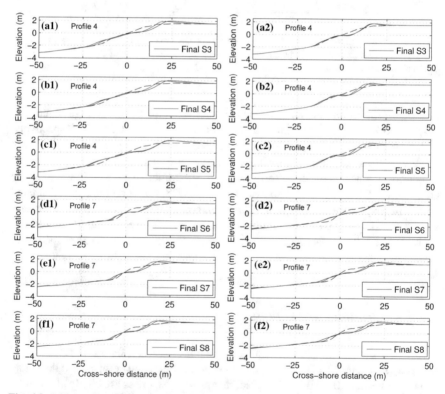

Fig. 6.9 Initial profile (black), final profile for scenario 0 (blue) and final profiles for the three scenarios with the highest morphological changes compared to scenario 0 under westerly (1) and easterly (2) storm conditions: **a–c** Profile 4 (Playa Granada), **d–f** Profile 7 (Poniente Beach). (*Source* [5]. Reproduced with permission of Elsevier)

These trends are summarized in Fig. 6.9, which depicts the morphological changes of two profiles (P4 and P7, Fig. 6.3) for the actual situation and for the three scenarios of wave farm location with the highest influence on the beach morphology (scenarios 3, 4 and 5 for profile P4, and scenarios 6, 7 and 8 for profile P7). It is observed that the highest decreases in beach erosion in the profile P4 (in Playa Granada) are obtained for scenarios 4 and 3 under westerly storms, and scenarios 5 and 4 under easterly storms (Fig. 6.9a–c). In the profile P7 (in Poniente Beach), the maximum protection is provided by scenarios 7 and 8: the reduction in beach erosion is higher for scenario 7 (scenario 8) under westerly (easterly) storm conditions (Fig. 6.9d–f).

The differences in bed level changes of all study profiles for each scenario with respect to scenario 0 under westerly storms are depicted in Fig. 6.10. It is observed that the greatest erosion reductions in Playa Granada (P3–P6, Fig. 6.3) are obtained with scenarios 3, 4 and 5; whereas scenarios 7 and 8 provide the greatest erosion decreases in Poniente Beach (P7 and P8, Fig. 6.3). The differences with respect to scenario 0 are significantly lower along the Salobreña section (P1 and P2, Fig. 6.3).

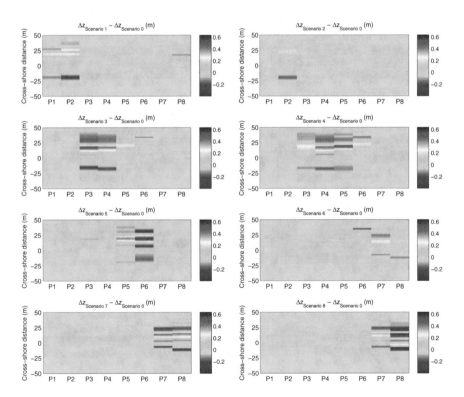

Fig. 6.10 Differences in bed level changes for each scenario compared to scenario 0 under westerly storm conditions in profiles P1 to P8. Positive (negative) values of the colour bar represent reduction (increase) in erosion with respect to scenario 0. Positive (negative) values of cross-shore distance correspond to the emerged (submerged) part of the beach profiles. (*Source* [5]. Reproduced with permission of Elsevier)

Altogether, the greatest protection along the entire deltaic coast under westerly storms is obtained by means of scenario 4, with an average decrease in the eroded volume equal to 14%. This value increases to 32% in the section most affected by erosion problems (Playa Granada).

Under easterly storm conditions, the greatest differences are obtained for scenarios 4, 5 and 6 (Fig. 6.11), with average reductions in the eroded volume up to 33% along the deltaic coastline and 70.6% in Playa Granada (both for scenario 5). Considering the weighted average of the values under both prevailing storm directions, the greatest protections are obtained for scenarios 4 and 5, with decreases of 23.3% and 22.8% along the entire coast, and 44.5% and 44% in Playa Granada, respectively (Table 6.2). These reductions in the eroded volumes contrast with the artificial nourishment projects performed at the study site since the entry into operation of the dam, with a total nourished volume of 250091 m^3 [9] and an average annual cost equal to 125045.5 euros/year [15].

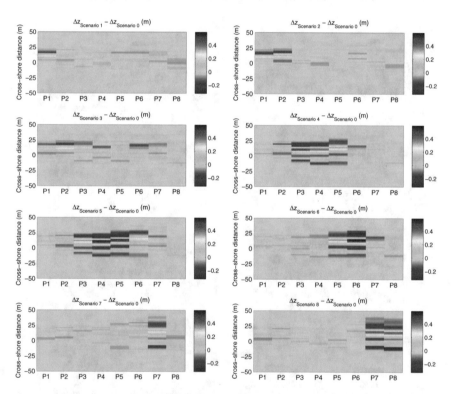

Fig. 6.11 Differences in bed level changes for each scenario compared to scenario 0 under easterly storm conditions in profiles P1 to P8. Positive (negative) values of the colour bar represent reduction (increase) in erosion with respect to scenario 0. Positive (negative) values of cross-shore distance correspond to the emerged (submerged) part of the beach profiles. (*Source* [5]. Reproduced with permission of Elsevier)

The obtained values of decrease in beach erosion are higher than those computed by [3], with overall reductions equal to 15% through the farm located at 25 m water depth and around 10% by means of the farms at 30 m and 35 m water depths. This higher reduction is attributable to the lower distance to the coastline in our study site (up to 1 km) with respect to the distances on Perranporth (about 2, 4 and 6 km for the farms located at water depths of 25, 30 and 35 m, respectively).

6.5 Conclusions

Although the influence of wave farms on the coastal dynamics has received increasing attention during recent years, the effects of farms on gravel-dominated deltaic beaches as a function of their alongshore locations have not been investigated. This chapter models the storm response of the Guadalfeo deltaic coast (southern Spain) for different alongshore locations of a wave farm through the joint application of a wave propagation model (Delft3D-Wave) and an extension of a morphodynamic model (XBeach-G) calibrated for the study site. The implications on wave propagation, run-up and beach morphology were quantified for each scenario of wave farm location and compared to the actual situation (scenario 0).

The influence of the different scenarios of wave farm location on wave propagation was studied by means of the quantification of significant wave heights in the leeward side of the farm (at 10 m water depth, H_{10m}). The lowest weighted values of H_{10m} along the entire deltaic coast and in the section most affected by coastline retreat (Playa Granada) were obtained with scenario 5 ($H_{10m} = 1.73$ m and $H_{10m} = 1.72$ m, respectively), followed by scenario 4 ($H_{10m} = 1.74$ m and $H_{10m} = 1.73$ m, respectively). The weighted reductions in H_{10m} for scenario 5 compared to the actual situation are equal to 9.8% and 18.3% in the whole deltaic coast and in Playa Granada, respectively (Table 6.2).

Regarding wave run-up, the effects of wave farm location were quantified by means of the 2% exceedence values of the wave run-up ($R_{2\%}$). The lowest weighted values of $R_{2\%}$ were again obtained through scenario 5 ($R_{2\%} = 1.09$ m along the entire deltaic coast and $R_{2\%} = 1.18$ m in Playa Granada) followed by scenario 4 ($R_{2\%} = 1.12$ m and $R_{2\%} = 1.2$ m, respectively). In this case, the decreases with respect to the actual situation are equal to 7.7% in the deltaic coast and 10.6% in Playa Granada for scenario 5, and 4.9% in the deltaic coast and 8.6% in Playa Granada for scenario 4 (Table 6.2).

To analyse the role of farm alongshore location in beach morphology, the storm-induced bed level changes were calculated in eight beach profiles. Under westerly storms, the greatest protection was obtained by means of scenario 4, with average decreases in beach erosion (compared to scenario 0) equal to 14% in the whole deltaic beach and 32% in Playa Granada. The maximum erosion reductions under easterly storms were obtained for scenario 5, with average differences respect to scenario 0 equal to 33% along the deltaic coast and 70.6% in Playa Granada. Considering both westerly and easterly storms, the reductions with scenario 4 (scenario 5) were equal to 23.3% (22.8%) in the deltaic coast and 44.5% (44%) in Playa Granada (Table 6.2).

According to the results obtained by [24], scenario 5 is the location with the highest values of wave power followed by scenario 6. They also found that storm conditions are not a limiting factor for the exploitation of WECs on the Spanish Mediterranean coast. These findings, along with the analysis of hydrodynamic and morphological variables presented in this chapter (summarized in Table 6.2), allows us to conclude that the best location in terms of both energetic performance and coastal protection is scenario 5.

The methodology followed in this chapter, based on the joint application of calibrated wave propagation and morphodynamic models, can be feasibly extended to other coasts across the world to select the best wave farm location in terms of reductions in wave height, run-up and/or beach erosion. This is particularly relevant in deltaic areas, subjected to increasing erosion problems due to human activities and specially vulnerable to the expected sea level rise in the coming years. In coastal regions with more energetic waves, complementary studies on the operational conditions and structural designs of the WEC farms would be required.

The analysis of the beach response during the lifetime of the wave farms (including non-storm conditions), along with the effects of the wave farms on the longshore sediment transport and coastline evolution, represent challenging research lines to be addressed in the future. This would require the joint application of wave climate simulations methods, Monte Carlo techniques, wave propagation toward the coast with downscaling methodologies, longshore sediment transport formulations and the one-line model over the lifetime period.

Acknowledgements This research was supported by the projects 917PTE0538 (CYTED - Programa Iberoamericano de Ciencia y Tecnología para el Desarrollo) and CTM2012-32439 (Secretaría de Estado de I+D+i, Spain), and the research group TEP-209 (Junta de Andalucía). The second author was funded by the TALENTIA Fellowship Programme (Regional Ministry of Economy, Innovation, Science and Employment, Andalusia, Spain) and the third author was funded by the Research Contract BES-2013-062617 (Spanish Ministry of Economy and Competitiveness). The authors are indebted to Alejandro López-Ruiz for his valuable suggestions and comments.

References

1. Abanades, J., Greaves, D., Iglesias, G.: Coastal defence through wave farms. Coast. Eng. **91**, 299–307 (2014)
2. Abanades, J., Greaves, D., Iglesias, G.: Wave farm impact on the beach profile: a case study. Coast. Eng. **86**, 36–44 (2014)
3. Abanades, J., Greaves, D., Iglesias, G.: Coastal defence using wave farms: the role of farm-to-coast distance. Renew. Energy **75**, 572–582 (2015)
4. Allen, J., Sampanis, K., Wan, J., Greaves, D., Miles, J., Iglesias, G.: Laboratory tests in the development of WaveCat. Sustainability **8**(12), 1339 (2016)
5. Bergillos, R.J., López-Ruiz, A., Medina-López, E., Moñino, A., Ortega-Sánchez, M.: The role of wave energy converter farms on coastal protection in eroding deltas, Guadalfeo, southern Spain. J. Clean. Prod. **171**, 356–367 (2018)

6. Bergillos, R.J., López-Ruiz, A., Ortega-Sánchez, M., Masselink, G., Losada, M.A.: Implications of delta retreat on wave propagation and longshore sediment transport-Guadalfeo case study (southern Spain). Mar. Geol. **382**, 1–16 (2016)
7. Bergillos, R.J., Masselink, G., McCall, R.T., Ortega-Sánchez, M.: Modelling overwash vulnerability along mixed sand-gravel coasts with XBeach-G: Case study of Playa Granada, southern Spain. In: Coastal Engineering Proceedings, vol. 1, Issue 35, p. 13 (2016)
8. Bergillos, R.J., Masselink, G., Ortega-Sánchez, M.: Coupling cross-shore and longshore sediment transport to model storm response along a mixed sand-gravel coast under varying wave directions. Coast. Eng. **129**, 93–104 (2017)
9. Bergillos, R.J., Ortega-Sánchez, M.: Assessing and mitigating the landscape effects of river damming on the Guadalfeo River delta, southern Spain. Landsc. Urban Plan. **165**, 117–129 (2017)
10. Bergillos, R.J., Ortega-Sánchez, M., Losada, M.A.: Foreshore evolution of a mixed sand and gravel beach: the case of Playa Granada (Southern Spain). In: Proceedings of the 8th Coastal Sediments. World Scientific (2015)
11. Bergillos, R.J., Ortega-Sánchez, M., Masselink, G., Losada, M.A.: Urban planning analysis of Mediterranean deltas-Guadalfeo case study. In: 12th International Conference on the Mediterranean Coastal Environment, vol. 1, pp. 143–154 (2015)
12. Bergillos, R.J., Ortega-Sánchez, M., Masselink, G., Losada, M.A.: Morpho-sedimentary dynamics of a micro-tidal mixed sand and gravel beach, Playa Granada, southern Spain. Mar. Geol. **379**, 28–38 (2016)
13. Bergillos, R.J., Rodríguez-Delgado, C., López-Ruiz, A., Millares, A., Ortega-Sánchez, M., Losada, M.A.: Recent human-induced coastal changes in the Guadalfeo river deltaic system (southern Spain). In: Proceedings of the 36th IAHR-International Association for Hydro-Environment Engineering and Research World Congress (2015). http://89.31.100.18/~iahrpapers/87178.pdf
14. Bergillos, R.J., Rodríguez-Delgado, C., Millares, A., Ortega-Sánchez, M., Losada, M.A.: Impact of river regulation on a Mediterranean delta: assessment of managed versus unmanaged scenarios. Water Resour Res **52**, 5132–5148 (2016)
15. Bergillos, R.J., Rodríguez-Delgado, C., Ortega-Sánchez, M.: Advances in management tools for modeling artificial nourishments in mixed beaches. J. Mar. Syst. **172**, 1–13 (2017)
16. Carballo, R., Iglesias, G.: Wave farm impact based on realistic wave-WEC interaction. Energy **51**, 216–229 (2013)
17. Deltares: XBeach-G GUI 1.0. User Manual. Delft, The Netherlands (2014)
18. Félix, A., Baquerizo, A., Santiago, J.M., Losada, M.A.: Coastal zone management with stochastic multi-criteria analysis. J. Environ. Manag. **112**, 252–266 (2012)
19. Fernandez, H., Iglesias, G., Carballo, R., Castro, A., Fraguela, J., Taveira-Pinto, F., Sanchez, M.: The new wave energy converter wavecat: concept and laboratory tests. Mar. Struct. **29**(1), 58–70 (2012)
20. Guanche, R., De Andres, A., Simal, P., Vidal, C., Losada, I.: Uncertainty analysis of wave energy farms financial indicators. Renew. Energy **68**, 570–580 (2014)
21. Holthuijsen, L., Booij, N., Ris, R.: A spectral wave model for the coastal zone. In: Ocean Wave Measurement and Analysis, pp. 630–641. ASCE (1993)
22. Lesser, G.R.: An approach to medium-term coastal morphological modeling. Ph.D. thesis, Department of Civil Engineering, Delft University of Technology, Delft, The Netherlands (2009)
23. Lesser, G.R., Roelvink, J.A., Van Kester, J.A.T.M., Stelling, G.S.: Development and validation of a three-dimensional morphological model. Coast. Eng. **51**(8), 883–915 (2004)
24. López-Ruiz, A., Bergillos, R.J., Ortega-Sánchez, M.: The importance of wave climate forecasting on the decision-making process for nearshore wave energy exploitation. Appl. Energy **182**, 191–203 (2016)
25. López-Ruiz, A., Bergillos, R.J., Raffo-Caballero, J.M., Ortega-Sánchez, M.: Towards an optimum design of wave energy converter arrays through an integrated approach of the life cycle performance and operational capacity. Appl. Energy, **under review** (2017)

26. López-Ruiz, A., Solari, S., Ortega-Sánchez, M., Losada, M.A.: A simple approximation for wave refraction-Application to the assessment of the nearshore wave directionality. Ocean Model. **96**, 324–333 (2015)
27. Losada, M.A., Baquerizo, A., Ortega-Sánchez, M., Ávila, A.: Coastal evolution, sea level, and assessment of intrinsic uncertainty. J. Coast. Res. 218–228 (2011)
28. Margheritini, L., Vicinanza, D., Frigaard, P.: SSG wave energy converter: design, reliability and hydraulic performance of an innovative overtopping device. Renew. Energy **34**(5), 1371–1380 (2009)
29. Masselink, G., McCall, R.T., Poate, T., Van Geer, P.: Modelling storm response on gravel beaches using XBeach-G. In: Proceedings of the Institution of Civil Engineers-Maritime Engineering, vol. 167 (4), pp. 173–191. Thomas Telford Ltd. (2014)
30. Masselink, G., Russell, P., Blenkinsopp, C., Turner, I.L.: Swash zone sediment transport, step dynamics and morphological response on a gravel beach. Mar. Geol. **274**(1), 50–68 (2010)
31. Matias, A., Williams, J.J., Masselink, G., Ferreira, Ó.: Overwash threshold for gravel barriers. Coast. Eng. **63**, 48–61 (2012)
32. McCall, R.T.: Process-based modelling of storm impacts on gravel coasts. Ph.D. thesis, Plymouth University, UK (2015)
33. McCall, R.T., Masselink, G., Poate, T.G., Roelvink, J.A., Almeida, L.P.: Modelling the morphodynamics of gravel beaches during storms with XBeach-G. Coast. Eng. **103**, 52–66 (2015)
34. McCall, R.T., Masselink, G., Poate, T.G., Roelvink, J.A., Almeida, L.P., Davidson, M., Russell, P.E.: Modelling storm hydrodynamics on gravel beaches with XBeach-G. Coast. Eng. **91**, 231–250 (2014)
35. Mendoza, E., Silva, R., Zanuttigh, B., Angelelli, E., Andersen, T.L., Martinelli, L., Nørgaard, J.Q.H., Ruol, P.: Beach response to wave energy converter farms acting as coastal defence. Coast. Eng. **87**, 97–111 (2014)
36. Millares, A., Polo, M.J., Moñino, A., Herrero, J., Losada, M.A.: Bedload dynamics and associated snowmelt influence in mountainous and semiarid alluvial rivers. Geomorphology **206**, 330–342 (2014)
37. Ortega-Sánchez, M., Losada, M.A., Baquerizo, A.: On the development of large-scale cuspate features on a semi-reflective beach: carchuna beach, southern Spain. Mar. Geol. **198**(3), 209–223 (2003)
38. Poate, T., Masselink, G., Davidson, M., McCall, R.T., Russell, P., Turner, I.: High frequency in-situ field measurements of morphological response on a fine gravel beach during energetic wave conditions. Mar. Geol. **342**, 1–13 (2013)
39. Roelvink D., Reniers, A., Van Dongeren, A., Van Thiel de Vries, J., Lescinski, J., McCall, R.: XBeach model description and manual. Delft University of Technology, User Manual, Delft, The Netherlands (2010)
40. Roelvink, D., Reniers, A.J.H.M., van Dongeren, A.P., van Thiel de Vries, J., McCall, R.T., Lescinski, J.: Modelling storm impacts on beaches, dunes and barrier islands. Coast. Eng. **56**(11), 1133–1152 (2009)
41. Rusu, E., Soares, C.G.: Coastal impact induced by a Pelamis wave farm operating in the Portuguese nearshore. Renew. Energy **58**, 34–49 (2013)
42. Stockdon, H.F., Holman, R.A., Howd, P.A., Sallenger Jr., A.H.: Empirical parameterization of setup, swash, and runup. Coast. Eng. **53**(7), 573–588 (2006)

Printed in the United States
By Bookmasters